D1163089

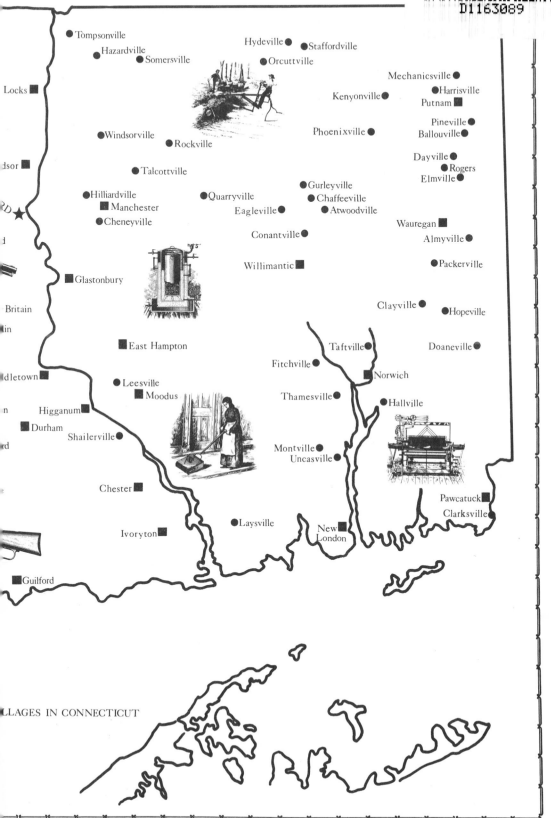

Locks ■

dsor ■

D ★

Tompsonville ●
Hazardville ●   Somersville ●
Hydeville ●   Staffordville ●
Orcuttville ●
Mechanicsville ●
Kenyonville ●   Harrisville ●
Putnam ■
Pineville ●
Phoenixville ●   Ballouville ●
Windsorville ●   Rockville ●
Dayville ●   Rogers ●
Elmville ●
Talcottville ●
Gurleyville ●
Hilliardville ●   Quarryville ●   Chaffeeville ●
Manchester ■   Eagleville ●   Atwoodville ●
Cheneyville ●
Wauregan ■
Conantville ●   Almyville ●
Willimantic ■   Packerville ●
Glastonbury ■
Clayville ●   Hopeville ●

Britain
lin

East Hampton ■
Taftville ●   Doaneville ●
Fitchville ●
dletown ■   Norwich ■
Leesville ●   Thamesville ●
Moodus ■   Hallville ●
Higganum ■
Durham ■
Shailerville ●
Montville ●
Uncasville ●
Chester ■
Pawcatuck ■
Laysville ●   Clarksville ■
Ivoryton ■   New London ■

Guilford ■

YANKEE DREAMERS AND DOERS

View of Colt Works, Hartford

Yankee Dreamers
AND
Doers

BY

ELLSWORTH STRONG GRANT

PEQUOT PRESS

Chester, Connecticut

Copyright © by Ellsworth S. Grant

ISBN: 87106-127-9
Library of Congress Catalog Card Number: 73-83256
Manufactured in the United States of America
All Rights Reserved
First Printing

TO
MARION HEPBURN GRANT

Who has always defended vigorously
those imperishable institutions of
family, community
and enterprise

Contents

List of Illustrations

TABLES

Preface

THOSE wise historians of the world, the Durants, summarized the basic lessons of history as being the laws of biology; competition, selection and fertility. "Competition", they said, "is the trade of life as it is the life of trade. Selection results from the natural inequality of man and repeatedly produces concentration of wealth, a peaceable or revolutionary redistribution and a renewed concentration. History, however, favors those nations in which the productivity of the land and the productivity of the women . . . in a word, fertility . . . is best in balance."

Certainly these laws are evident in the growth of Connecticut industry, especially in the decades from 1790 to 1860, and this is what this book is about. The merchants and manufacturers, the artisans and machinists, the peddlers and tinkers all demonstrated the revolutionary nature of the American system. They did not defend the status quo; they assaulted it both technically and socially. Many concepts of business management and social organization are actually the offspring of Connecticut Yankee pioneering.

Seldom do new ideas or technologies emerge without a foundation in old theories and experiments. This era from 1790 to 1860 was the heroic, eruptive period of capitalism . . . vibrant, experimental, individualistic, crude and sometimes cruel . . . far removed from our current era of corporate empires and colossi.

Here, then, in this book, is the background to the development of today's unique American economic system. To understand it, to enlarge upon it, may be timely hindsight; for, as Carl Sandburg said: "When a nation or society perishes, one condition may always be found: they forgot where they came from." The novelist, Robert Penn Warren, similarly observed that in creating an image of the past we are creating ourselves and without it we do not exist.

Hopefully, this book will serve to create this vital image of the past for the younger generation, of how and why their predecessors created our industrially oriented culture, the golden cornucopia and monster that has produced so much for so many.

". . . There are *Two Callings* to be minded by *All Christians*. Every *Christian* hath a GENERAL CALLING; which is, to Serve the Lord. . . . But then, every Christian hath also a PERSONAL CALLING; or a certain *Particular Employment*, by which his *Usefulness*, in his Neighborhood, is distinguished . . . There should be some *Special* Business . . . wherein a Christian should for the most part spend the most of his Time; and this, that so he may Glorify God, by doing of *Good* for *others*, and getting of *Good* for himself . . . Yea, a *Calling* is not only our *Duty*, but also our *Safety* . . . A Christian should follow his *Occupation* with INDUSTRY . . . DISCRETION . . . HONESTY . . . CONTENTMENT."

COTTON MATHER

CHAPTER I

A Time for Dreaming and Doing

AMERICA'S Revolution had ended. Freedom having been won, the time came to lay the foundation for a new political and economic system, different from anything the world had ever seen. At the Constitutional Convention in Philadelphia, for a long, hot summer of debate, gathered the delegates of the thirteen states in 1787. Connecticut sent the Windsor lawyer, Oliver Ellsworth; the modest, scholarly president of Columbia College, William S. Johnson; and the old, loquacious, awkward politician, Roger Sherman. Thomas Jefferson claimed that Sherman never said a foolish thing in his life, but the illustrious Hartford merchant, Jeremiah Wadsworth, who knew him better, confided that he was as "cunning as the Devil." Yet Sherman and his associates had the temerity to assert that they represented "a manufacturing state"! "Already," Ellsworth proclaimed, "it makes its own tools of husbandry and half its own clothing."

Yankees, of course, have never been known to hide their lights under a bushel, but at this time the little rectangle on the map called Connecticut contained fewer than 240,000 people. It had only a relatively tiny external trade, small deposits of copper and iron ore, poor roads, and barely enough tillable land to support its population, the younger ones of which were already emigrating north and west. A few shipyards flourished in the river ports and along the Sound, providing bottoms made from local hardwoods for the coastal and Caribbean trade. The hard-drinking colonists regarded old "kill-devil" rum, the number one export to Europe, as a necessity for survival.

During the rebellion, a handful of infant industries dotted the state. In Norwich, Christopher Leffingwell made woolen cloth; in Middletown, George Starr had a shoe and leather goods shop. Powder mills sprang up in several towns, one of the best-known run by the enterprising East Hartford Pitkin family, who were also engaged in making glass, guns, anchors, snuff, felt and cotton stockings. Gunsmiths were offered a state bounty to increase their output of flintlocks. Owing to its high-grade iron ore, Salisbury foundries became the

Old Bidwell Mill, Oxford

cannon center of New England, giving the state its long-standing reputation for being "the arsenal of democracy." There even existed a surplus of the famous Yankee notions. Isolated as they were, the colonists had succeeded in becoming independent enough to take care of their own needs, but not so much as to offer real competition with England.

What Connecticut's delegates undoubtedly intended by their dramatic boast was to publicize the Yankee's dogged self-sufficiency, his penchant for wearing out, using up, making do, and doing without, his ingenuity with farm tools and contraptions, his love of tinkering with wood and iron. In winter, every farmhouse had to be a domestic workshop, with a spinning wheel and maybe a loom. Grist, saw and fulling mills, as well as blacksmith shops, breweries and tanneries, were common to most villages. They processed grain, wood, leather and cloth, and fabricated iron, but for local consumption only. Until the introduction of the marvelous new machines being invented in England and the widespread use of water power, these could not aspire to be more than simple shops, using the crudest of equipment and employing, besides the proprietor, one or two part-time helpers. Clever artisans turned out tin pots and pans, shoes, hats, buttons and wooden clocks, but in limited quantity. In the sense of having factories, with labor-saving machines, production lines and wage earners, the pronouncement of the three Connecticut delegates amounted to only a prophecy of events to come, inspired to be sure, but then scarcely defensible.

Water wheel, New London, 1650

* * *

Beneath the Connecticut Yankee's genius for inventing, manufacturing and peddling lay not only a tradition of frugality and independence, but, generally speaking, an enduring faith that motivated him as compellingly as it had his forebears who sailed to Plymouth. In John Chamberlain's succinct phrase, he was "inherently a self-starter," who saw no irreconcilable conflict between faith on the one hand and enterprise on the other. In fact, the concept of the free enterprise or incentive system, at least in embryo, came to this continent with the Pilgrims and later settlers. Self-interest, as much as religious faith, was behind emigration. The ships that carried the colonists were owned by joint stock companies and merchant adventurers who hoped to profit from the trading activities of the settlers. The Massachusetts Bay Colony was founded by a company headed by an English lawyer, John Winthrop, and supported by a royal charter and the credit extended by London merchants—in other words, a business venture from the beginning.

Between what the eminent German sociologist, Max Weber, called the "Protestant Ethic" and capitalism, there is a close relationship. Weber traced the origin of the ethic to the religious writings of John Calvin, who claimed that God had predetermined man's fate. It was therefore natural to assume that success in one's calling or occupation would be an indication of having been chosen to be saved. In the early 1600s John Smith and John Winthrop clearly espoused the necessity of work for survival in the harsh New World. At the beginning of the eighteenth century, Cotton Mather elaborated on this concept. In extraordinarily clear and simple terms he delineated the uniquely American union of religion with enterprise. Every Christian, he preached, has two callings of equal importance: one to save his own soul by serving the Lord, the other to pursue an occupation in order to do good to others as well as to acquire good for himself. A Christian, he said, should work with industry, honesty, discretion and contentment. The work ethic, as defined by Mather, did not operate without conflict between Puritan ideology and individual economic behavior. When merchants, for example, found it difficult to strike a balance between service to God and Mammon, they invariably turned away from the restrictions of Puritanism. By the mid-nineteenth century most New England Yankees were practicing their personal calling so assiduously that the religious one had been virtually abandoned.

As it developed, the Anglo-Saxon approach to economic life comprised six main ingredients: political freedom, importance of the individual, competition, practicality or common sense, optimism and the moral aim to advance the welfare of all. From the beginning, the exigencies of frontier living and the isolation of the colonies also encouraged self-reliance. In his penetrating evaluation of American life, de Tocqueville theorized that democracy had the effect of diverting the farmer from his land and stimulating his taste for commerce and manufacturing because of a natural desire for "physical gratification" and also "for the love of the constant excitement" of business activity. "But what most astonishes me in the United States," he added, "is not so much the marvelous grandeur of some undertakings as the innumerable multitude of small ones."

Benjamin Franklin's "poor Richard" epitomized the shrewd, industrious, self-reliant businessman of the late eighteenth century who wanted to be left alone to find his salvation. In fact, Franklin was the first to use the term *laissez faire* in English when he published, in 1774, his *Principles of Trade*, a work derived from the writings of French physiocrats like Mirabeau and duPont de Nemours. Adam Smith's *The Wealth of Nations* in 1776 provided the rebels with the rationale for free enterprise. Like the physiocrats, he advocated "a system of natural liberty," yet favored restraint of individual actions that might endanger the security of the whole society. His views amounted to a condemnation of an English mercantilism that, as he saw it, conferred monopolistic advantages upon a handful and severely restricted individual freedom. His book, which went through three editions before 1820, became the gospel of economic life for Americans. "Adam Smith," according to Nye, "became by adoption an American Founding Father. Few men have had a more powerful influence on American life." Between Cotton Mather's "personal calling" and Adam Smith's "natural liberty," therefore, a strong belief evolved in the worth of the individual and the righteousness of free enterprise. That such a combination of religion and philosophy would eventually degenerate into a cult of materialism and exploitation in much of the America we know today was certainly not foreseen by the optimistic founders and pioneers of our industrial society. This psychological conditioning, reinforced by the physical circumstances of being colonists ruled by a misguided, absentee crown, had to culminate in the struggle for first political and then economic independence.

Although the freedom-seeking, get-up-and-go spirit of the Yankee was chiefly responsible for the rapid development of manufacturing during the first decades of the nineteenth century, he nevertheless owed a great deal to the technical progress achieved in the mother country. The jealously guarded ideas of the pioneering English mechanics who sparked the Industrial Revolution had to reach America through gifted *émigrés* in order for the new nation to convert from homespun to factory. The first of the great textile innovations that revolutionized Great Britain toward the end of the eighteenth century was James Hargreaves's spinning machine, containing multiple spindles activated by one wheel. He called it a "jenny" after his wife, and by this name it has been known ever since. In 1769, Sir Richard Arkwright, one of those few inventors in the industrial age who were able to accumulate a fortune as well as fame, patented a spinning frame and at Nottingham set up the world's first cotton mill operated by horsepower. A few years later, Samuel Crompton, seeking a method to handle the finest yarns that would combine the features developed by Hargreaves and Arkwright, came up with his "mule." From the spinning of yarn, English inventors turned their thoughts to making a power loom for weaving, a device that sprang from the mind of a nonmechanical minister named Edmund Cartwright in 1785.

Before cast iron became plentiful, England's textile mills depended upon wooden machines, gears, shafts, and pulleys. Even the early lathe was made mostly of wood. Not until the ironmaster John Wilkinson, just before the American Revolution, produced a machine for boring cannon, did England have the first metal-cutting tool capable of doing large work with anything like modern precision. Its effect was revolutionary. With his boring miller, any number of cylinders could be made so nearly perfect that they would not err "the thickness of an old shilling in any part." Its appearance also illustrates the interdependence of mechanical inventions because James Watt could never have finished his famous steam engine, which, in 1775, Wilkinson ordered for his ironworks, without the latter's ability to furnish him accurate metal cylinders. Twenty years passed before Henry Maudslay, who is regarded as the father of machine tools, made the forerunner of the modern lathe and slide rest, with the help of the inventive brilliance of Samuel Bentham and Marc Brunel. From Maudslay's lathe spawned the whole family of machine tools—the miller, jig borer, grinder and screw machine, more than 250 different kinds today—that gave rise to the great metal-working industries of the nineteenth and twentieth centuries and accounted for a large measure of the world's technological miracles.

Determined to keep its American colonies dependent on the mother country, at least for manufactured goods, Britain did its best to prevent native enterprise from prospering. Most of the copper from East Granby's Newgate mines had to be exported rather than converted for domestic use. The emigration of mechanics was strictly prohibited in order to keep English production methods at home and the American worker unskilled. Yet, after the Revolution enough able, energetic men did manage to find their way to America to get such industries as cotton and brass started. One was Thomas Marshall, who had worked for Arkwright and applied to Alexander Hamilton for a position in the latter's experimental cotton mill in Paterson, New Jersey. Another was Samuel Slater, who in 1790 recreated from memory Arkwright's spinning machine in the Pawtucket mill of Moses Brown. President Jackson subsequently praised Slater as "the father of American manufactures."

The history of Pawtucket's prolific and gifted Wilkinson family further illustrates the close interrelationship between English and early American inventiveness. Oziel Wilkinson, Slater's father-in-law but no relation to the John Wilkinson of boring fame, was a partner with Slater in building his second cotton mill in 1798. Oziel also ran an anchor forge, where his son David—in one of those rare instances of simultaneous discovery—created a lathe with a slide rest and lead screw at the same time as Maudslay. In 1810, David converted the family forge into a shop for fabricating cotton machinery, the demand for which followed hard on Slater's successful venture.

In a move to become independent of Slater, Wilkinson's father erected a factory in Pomfret, Connecticut, for spinning cotton yarn, an investment of $60,000. For this mill on the Quinebaug River, David supplied his first engine lathe. He also perfected the power loom for cotton. Although Francis Lowell pioneered power weaving at the Waltham mill of the Boston Manufacturing Company in 1814, Wilkinson's loom design was the one that came into general use throughout the United States. Fifty years after he obtained his lathe patent, in 1847, Congress granted him $10,000 for his achievement. At that time over 200 Wilkinson lathes were being operated in U.S. arsenals.

* * *

Connecticut's most abundant resource was water for power. Until the 1840s, the availability of water determined the location of most mills and factory villages. In colonial times, a miller usually built a tiny dam and canal on a small stream and erected a wooden undershot waterwheel, a separate one being necessary for each machine. His neighbors brought grain to the gristmill, or cloth woven during the winter months for finishing in the fulling mill. The first known dam in the Connecticut Valley, one of stout oak, was built by Leonard Chester for a gristmill in Wethersfield as early as 1637. The mill became a community center, where colonists gathered not only for flour, but to socialize and to hear the news.

Mill rights soon became foundation stones in New England's economic structure and formed an important part of riparian law. Dams not only stored surplus water, but also stabilized the stream flow and kept the mill turning day and night. At the falls on the Yantic River in Norwich, for example, America's first linseed oil mill started in 1718; at the same location, in 1766, appeared a paper mill. Its owner, the prominent merchant, Christopher Leffingwell, wrote in 1791:

> The Stream which runs thro this Town has many Mill Seats more & better than in any Other in this State and a great Variety of Manufacturing will be done by Machinery turned by Water.

Leffingwell employed about a dozen men who produced 1,300 reams of paper per year. At the same time he ran a stocking mill. The country's first woolen mill, in 1788, depended upon the Little River in the center of Hartford for waterpower. And when Eli Whitney, in 1798, built the first real firearms factory, as well as the first factory village, he chose for his location the falls on Mill River in Hamden, on the outskirts of New Haven.

Not all villagers welcomed the building of dams and the erection of mills for fear that America would duplicate the evils of the dismal factory system. Covertly bringing cloth designs from England, twenty-three-year-old John Warburton started a cotton mill in Connecticut as early as 1795. He chose Vernon for his site and Samuel Pitkin for his partner. The neighbors did not believe he could possibly succeed and refused to offer any help. Warburton himself had to move the gravel for his dam. Unable to rent a house, he lived in his mill. Obtaining labor was his next vexing problem. Finding no one willing to spin cotton for his stocking yarn, he appealed to his wife. But she, poor woman, was busy all day caring for their infant children. Undaunted, he solved the problem by means of a large cradle and a mechanical arrangement for rocking it that intrigued the neighborhood. Both cradle and mill were immediate successes.

Goodwin's Paper Mill

Goodwin's Paper Mill (interior)

Apparently, however, good fortune went to John's head. He built a big brick house with four chimneys, which later became the Warburton Inn. Convivial by nature, he kept a hogshead of Jamaica rum on tap in an open shed near the road, free to all comers. In the end, his liberal hospitality and fondness for "kill-devil" ruined both his health and his business.

* * *

America's destiny as either an industrial or agricultural society was hotly debated after the Revolution. Thomas Jefferson's dream of a pastoral paradise appealed to those who had fled religious persecution in Europe, been displaced as tenants by the rush to industrialism in England or slaved in English cotton mills. John Adams told Benjamin Franklin in 1780 that America would not make manufactures enough for her own consumption these thousand years. In 1782, Jefferson argued that his country did not require economic self-sufficiency because of its surplus of land. He wrote: "Let our workshops remain in Europe. It is better to carry provisions and materials to workmen there, than bring them to the provisions and materials, and with them their manners and principles . . . I consider the class of artificers as the panders of vice . . ." Agriculture, he said, was the only true source of wealth. Yet, three years later, aware of the American taste for navigation and commerce, he began to see the impracticality of his theory of a rural society and even recognized that the new-fangled steam engine would someday be applied generally to machines.

Secretary of the Treasury Alexander Hamilton's *Report on the Subject of Manufactures*, published in 1791, turned the tide in favor of industry. In part a census of the infant status of manufacturing before 1800, it listed seventeen different kinds of on-going enterprise besides household spinning and weaving, including ironworking, tanning, sugar refining, shipbuilding, tinware, copper and brass work and hatmaking. Its philosophy reflected Adam Smith's *Wealth of Nations* and constituted, according to Arthur Scheslinger Jr., "the first great expression of the industrial vision of the American future." Accepting Jefferson's faith in a predominantly agricultural society, at least for the purpose of his thesis, Hamilton, nonetheless, made a case for government support of manufactures on the basis that they would advance, rather than injure, the real interests of farming. Manufacturing, he argued with forthright logic, would promote the division of labor, skill and dexterity, use of machinery, emigration from abroad, employment and demand for farm products. His conclusion was a plea for its protection through duties and bounties. The report brought no immediate action, yet it served as a reservoir of argument and inspiration for later generations.

During his presidency, Jefferson himself was ready to abandon his own dream. By 1809, he admitted that an equilibrium of agriculture, manufacturing

Arkwright's Jenny, 1775

Site of revolutionary foundry, Salisbury

and commerce were essential, and when the first serious effort came in 1816 to do something about Hamilton's report, he remarked contritely: "Experience has taught me that manufactures are now as necessary to our independence as to our comfort." Yet he never gave up hope that industry would remain in the household.

Another eager advocate of manufacturing was Tench Coxe of Philadelphia. The young, controversial and ambitious Coxe had as keen an understanding of economics, as clear a foresight of the new nation's industrial destiny, as any American then alive. Coxe worked closely with Hamilton in drawing up his report. To the delegates at the Constitutional Convention, he talked up the need for a balanced economy, and especially for domestic industry as a convenient market for the surplus produce of the soil. Coxe was not too far ahead of his time when he suggested that the high cost of labor, lack of skilled hands and scarcity of materials could be overcome by the machine: "factories which can be carried on by water-mills, windmills, fire, horses and machines ingeniously contrived . . ." Two years before Slater's arrival, he unsuccessfully tried to have models of Arkwright's spinning frame imported. Moreover, citing the example set by Slater and his Sunday school for child laborers, Coxe was equally optimistic that America would purify the factory system.

While the Constitution was being ratified, there occurred close to Philadelphia two major technological achievements that gave substantial weight to Coxe's arguments. John Fitch, a native of South Windsor, Connecticut, demonstrated the practicality of applying steam to the propulsion of a vessel on the Delaware River. Oliver Ellsworth was one of his passengers. And America's first great engineer, Oliver Evans, perfected an automated production line for the milling of grain, devising a system of conveyers driven by the same waterwheel that turned the grinding stones. Only two men were needed in his mill, one to pour in the wheat from sacks at one end, another to nail up the barrels of flour at the other. Later Evans introduced the high pressure steam engine, which eventually enabled industry to convert from water to steam power.

Ratification of the Constitution in 1788 was in itself a powerful stimulus to enterprise by creating free trade between the states and giving rise to the greatest common market in history. In the same year, three companies shared the honor of being the first textile mills in the United States. All of them used the spinning jenny. One was the Pennsylvania Society for the Encouragement of Manufactures & the Useful Arts, with Coxe as president. At Beverly, Massachusetts, three Scotsmen launched a cotton mill, a timid venture that wove corduroys and bed ticking for export and survived only until the Embargo of 1807. And in May, the Hartford Woolen Manufactory commenced business.

This event was an important milestone in another way, because it anticipated the transformation twenty-five years later of merchant capitalists into industrialists. The power structure of this period, appropriately called the "Standing Order," consisted of a tightly-knit, self-perpetuating group of ministers, lawyers and merchants from the leading families who controlled the affairs of church and state well into the next century. They were solid Congregationalists and staunch Federalists, whose meeting houses served as both church and town hall. The fortunes accumulated by Connecticut merchants may not have been as impressive as those of the great Massachusetts shipping interests, with their larger ships and ocean ports, but in relation to population there were more of them.

By far the most outstanding, Jeremiah Wadsworth of Hartford stood shoulder high with the leading men of commerce in Boston and New York, and on a nearly equal footing with Hamilton, Washington and Lafayette. A minister's son, bound out as an apprentice at fourteen, he became a ship's captain in the West Indian trade during his twenties, learning to buy and sell cargo at the right price. By the age of thirty he owned at least one brig. During the Revolution, as Commissary General of Purchases, without salary, for both Washington's army and the French forces, he saw no reason why patriotism should interfere with profit. Along with General Nathanael Greene, then Quartermaster General, he benefited as a silent partner in the mercantile establishment of Barnabas and Silas Deane of Wethersfield, who provided staples and manufactures for the Army and operated several rum distilleries. The wealthiest man in Connecticut at the end of the war, Wadsworth also speculated in credit, founded Hartford's first bank in 1792, and even served six years in Congress. He was a risk taker without peer.

Wadsworth's participation in the woolen mill was the height of speculation. Casting about for new forms of investment with their surplus funds, Wadsworth and the other founders decided to see whether American-made wool could compete with the popular British imports. Why not put a dollar or two down and spin the wheel of fortune? In ten-pound shares they raised £1,280 and in January, 1789, despite inferior equipment and almost complete ignorance of textile methods, managed to ship cloth to New York for sale.

Among the thirty-one stockholders were such important personages, besides Wadsworth, as Peter and Elisha Colt, the jurist Oliver Ellsworth and Governor Oliver Wolcott, Sr. A Yale graduate and successful merchant, Peter Colt, son of Deacon Benjamin Colt of Lyme and the great uncle of Samuel Colt, had been deputy commissioner general of purchases for the Eastern Department (New England and New York) under Wadsowrth. From 1789 to 1793 he served as state treasurer and then moved to Paterson, New Jersey, to take charge of the cotton mill in which Alexander Hamilton was so interested.

Jeremiah Wadsworth and his son Daniel

Elisha Colt, Peter's nephew, whose father was an axe and scythe maker in Lyme, seems to have been the mill's manager, as well as responsible for marketing its output. In 1819, this popular and efficient man became the first treasurer of Society for Savings, the sixth oldest mutual savings institution in the country, at the same time doubling as Connecticut's treasurer. His daughter married Samuel Collins, the founder of the famous axe company.

For Hamilton's survey of American industry, Elisha Colt described in detail the mill's operation:

> We have now erected a Building . . . in which we have three Broad and five Narrow Looms at work—which will consume about 12,000 pounds of Wool p annum—Our Wool is collected about the Country from the Farmers who raise it, it costs from ½ to ⅙ (shillings) the pound, taken at their Houses—when it comes into the Wool store, the Fleeces are broken up and sorted exactly in the manner practised in England, by workmen regularly bred to that branch . . .

After being cleaned and oiled, the wool was turned over to "families in the neighboring districts to be spun into yarn—when it is returned to the Factory and there warped into proper pieces and wove into Cloth in our own Looms . . . from the Loom they go to the fulling Mill." The first year's production totaled 5,000 yards, some of which sold for as high as five dollars a yard.

According to Colt:

> the Embarrassments we labour under arise from the Scarcity of Wool, and the consequent high price of that article; and the frauds and impositions we are subject to in purchasing Wool, from a total want of inspection or regulations in packing the Wool for market—from the scarcity of Workmen . . . from the want of machines for expediting Labour; such as Scribling or Carding machines, that are worked by Water, and Jennies for spinning yarn—
>
> The present use of Machines in England give their Manufacturers immense advantages over us—This we expect soon to remedy—But the price of Wool must be high with us, until the people of the middle & southern States can be induced to turn their attention to raising Sheep—should this take place, in a very few years we may raise wool enough for the consumption of this Country & make every kind of Woolen Goods suitable for every description of People from the highest Class of Citizens to the Negro Slaves . . .

Although the company never sought a charter, the legislature seemed to appreciate the uniqueness of this venture, and in a rare display of tangible support for domestic industry thrice heeded its appeals. First, by an act of May, 1788, a bounty of one pence per pound was offered on yarn spun and made into cloth for one year; the following year it voted to exempt the Hartford mill and three others—in Farmington, New Haven and Killingly—from taxes for a period of five years and relieved their employees from paying poll taxes for two years. In 1790, the directors petitioned for direct financial aid because of declining sales and their inability to purchase enough raw material to keep going. Citing New York's precedent of making a grant for the promotion of linen manufacturing, they asked for £750. The request was rejected, but the fall session of the legislature gave them permission to conduct a public lottery to raise £1,000. Elisha Colt convinced the legislators that since the invested capital had been depleted even after increasing it to £2,800, a lottery was the only way the infant establishment could hope to meet British wool competition.

Defined by one wag as "the only method by which laziness and greed could be combined to satisfy both," state-approved lotteries had for a long time been common in New England as the one approved form of gambling. After a fire Fanueil Hall in Boston was rebuilt by lottery in 1761; the Beverly Cotton Manufactory benefited from a Massachusetts state lottery in 1791; and the State House in Hartford depended upon a lottery for its construction in 1796. Most lotteries were limited in size, realizing not more than $1,000 or $2,000 in profit. After the turn of the century, however, they lost favor with the legislators, especially when their multiplicity resulted in poor ticket sales and insufficient net returns.

For the Hartford mill Elisha Colt netted $3,000 from the authorized lottery, twice the amount he anticipated, which he used to purchase new equipment, including a much-wanted spinning jenny. Since England forbade the exportation of wool machinery, the mill had to improvise satisfactory substitutes, probably relying on locally made machines. After 1792, power-driven machinery was introduced, and yarn spinning was brought from the homes into the mill. The jenny, however, an adaptation of the English cotton frame, was worked by hand. Even with the infusion of additional funds, the mill's future seemed dubious. Peter Colt wrote: "Those persons concerned in setting up new Manufactures have every obstacle to surmount which can arise from clashing Interests, or ancient prejudices; as well as from the smallness of our capitals, the scarcity of Materials & workmen, & the consequent high prices of both . . . Some kind of aid therefore, from the General Government of the United States will be necessary in order fully to establish Manufactures . . ." Although the partners found a source of skilled workmen among British deserters and prisoners of war who had decided to remain in the new country, these proved unreliable and intractable. Those few citizens who could afford to buy woolens still preferred British-made goods, which, of course, delighted the English agents who were going all out to prevent the rise of any domestic industry.

But no less a personality than our first President gave the mill a much-needed boost. At his inauguration in January, 1789, General Washington

SCHEME OF A LOTTERY,

For the purpose of extending and improving the Woollen Manufactory in the City of Hartford, agreeably to an Act of the Legislature of the State of Connecticut, passed October 1790.

CLASS THE FIRST.

Prizes.		Dollars.		Dollars.
1	of	1000	is	1000
1		500		500
2		300	are	600
2		200		400
6		100		600
10		50		500
15		40		600
20		30		600
30		20		600
120		10		1200
2850		4		11400

3057 Prizes. ⎫
5943 Blanks. ⎬ 9000 Tickets at 2 Dol. each, 18000

NOT TWO BLANKS TO A PRIZE.

Subject to a Deduction of Twelve and Half per Cent.

THE object of this Lottery is to enable the Proprietors of the Hartford Woollen Manufactory to procure Machines, Implements, and to increase their Stock, in order to render the Business more extensively useful to the Community; and the Managers flatter themselves that the Scheme is calculated much to the Advantage of Adventurers who wish to put themselves in FORTUNE's Way, as well as to those who are disposed to buy on patriotic Motives.

The Drawing is proposed to begin by the first Day of February next, or sooner if the Tickets are disposed of. A List of the fortunate Numbers will be published in the Connecticut Courant, and Prizes paid on Demand by the Managers. Those Prizes not called for in six Months after the Drawing, will be deemed as generously given for the Use of the Factory, and appropriated accordingly.

HEZEKIAH MERRILL, ⎫
ANDREW KINGSBURY, ⎬ Managers.
ELISHA COLT, ⎭
Hartford, November 1790.

Hartford Woolen Manufactory's lottery scheme

WANTED at the Woolen Manufacture feveral fprightly, active Lads, (as Apprentices) from 14 to 16 years old— of good characters and abilities—care will be taken to inftruct them in fuch branches of the bufinefs, as fhall be moft agreeable to them.—A good narrow or Broad-Cloth Weaver, and one or two good Scriblers, would find conftant imployment, and good wages—-by applying to DANIEL HINSDALE.

☞ CASH paid for WOOL.
Hartford, July, 1788.

Advertisement for apprentices in *Connecticut Courant*

wore a dark-brown suit, as did Vice-President John Adams, made from thirty yards of Hartford's best broadcloth. Colonel Wadsworth, who made the presentation, thoughtfully brought along a similar amount of cloth for the first First Lady. The Prussian Inspector General "Baron" von Steuben, who also wore a suit of Hartford wool, invented for it a button made from Indian wampum. Senator Oliver Ellsworth, Wadsworth and the other Connecticut representatives elected to Congress were similarly attired. After a visit to the Hartford Woolen Manufactory the following October, the President wrote in his diary a somewhat qualified endorsement: "Their broadcloths are not of the first quality as yet, but they are good . . . I ordered a suit to be sent to me at New York." Upon receiving it, he told General Knox that it exceeded his expectations. One product, the "Hartford Gray," became quite popular, almost matching English fabrics in quality and finish.

Despite payment of a fifty percent dividend in finished goods at the end of 1794, sales were still disappointing. Several times the inventory had to be disposed of at auction. An English clothier found the mill in decay, the machinery inadequate, the cloth poor. He reported: "$9,300 have been lent towards the undertaking by the State. None of the partners understand anything about it and all depends on an Englishman who is a sorter of the wool." The following August the little mill closed after six and one-half years of operation. At an auction Colonel Wadsworth bought the property, which included, besides 140 pieces of cloth, 2 carding machines, 8 looms, a twisting machine and the jenny.

In his Report Hamilton recognized the significance of this noble, but ill-fated, experiment:

> Their quality certainly surpasses anything that could have been looked for in so short a time, and under so great disadvantages . . . To cherish and bring to maturity this precious embryo, must engage the most ardent wishes . . .

Unfortunately, the embryo was too premature to survive. For one reason it lacked an adequate market. If Wadsworth and the Colts could have prolonged their venture a few years, they might have been able to take advantage of the higher-quality wool from the domestic breeding of Merino sheep, the development of improved machines operated by waterpower, and the growing inclination to favor American-made products. Yet, as the dean of the wool-industry historians, Arthur H. Cole, stated, it was "the first purely wool-manufacturing concern founded on a strictly business basis, and the first in which power machinery was employed . . ." for even one or two operations.

*   *   *

To assemble the data for his *Report on Manufactures*, Hamilton sent a circular letter to the supervisors and inspectors in the thirteen states who were appointed to collect the internal revenue duties which had been imposed in March, 1791. In Connecticut, John Chester of Wethersfield was the principal agent. He consulted as many well-informed businessmen, state officials and legislators as he could reach. The informants were extremely candid and detailed in their responses, as typified by Elisha Colt's description of the troubles of the Hartford woolen mill. Some manufacturers even went to the bother of forwarding samples of their work. Colt supplied woolen cloth, John Mix of New Haven some of his metal buttons and the Town of Mansfield its sewing silk. Common to these reports were a desire to nurture domestic industries as an antidote to imports, and a plea for federal assistance. Mr. Learned in New London quoted manufacturers as believing that tariff duties on saddles were not high enough; John Mix wanted foreign-made buttons banned; Peter Colt expressed a hope that the country be freed from dependence on Europe for ordinary clothing. John Treadwell of Farmington referred to the difficulty of marketing; a local textile producer had to dispose of his goods through barter because of inferior quality, while cash for imports was plentiful. Hatmaking in Danbury suffered from persons entering the business who never served an apprenticeship in the trade. And Peter Colt decried the indulgence in paper speculations.

According to Colt, the manufactures of Connecticut were divided as follows:

> Those carried on in Families merely for the consumption of those Families;—those

carried on in like manner for the purpose of barter or sale; & those carried on by tradesmen, single persons, or Companies for Supplying the wants of others, or for the general purposes of merchandize, or Commerce.

Those . . . which are purely domestic, are the most exclusive and important; there being scarcely a Family in the state either so rich or poor as not to be concerned there in. These domestic Manufactures are of Linen, of Cotton & of Wool . . .

From these raw materials, he said, were made hose, tow cloth, coarse linens, shirts, bedticks, also "coarse fustians and Jeenes for mens wear, & white Dimity of the Women," woolens, thread and sewing silk. In addition, he listed products of wood, iron and leather, both for home consumption and exportation, such as household furniture, carriages, cutlery, tools, shoes, saddles, bridles and horse harness. In a postscript he added "tin men, pewterers, Hatters &c. & Silversmith in a great plenty—Braizers, Brass founders,— & of late Button makers. . . ." He did not dare estimate the value of all these manufactures, but noted that it was considerable and increasing yearly.

Christopher Leffingwell advised Chester that, in his opinion, Norwich, being at the head of a navigable river and having access to many mill seats, would become an important manufacturing center, especially for cotton goods, with the carding, spinning and weaving all done by means of waterpowered machines.

Danbury, for its part, was already gaining a reputation for hats. Joseph P. Cooke, a wealthy merchant, state legislator and Congressman, wrote Chester:

The manufacturing of Hats of all kinds is prosecuted upon a large scale in this town; from the factory of O. Burr and Co. which is probably the largest of its kind in this State, large quantities of hats are sent abroad . . .

Hatmaking was fairly widespread in western Connecticut during most of the eighteenth century, although Zadoc Benedict is often credited with being the first to have a shop. He started in 1780 with a single kettle and the labor of three men, who produced eighteen hats a week. O. Burr & Company started in business in January, 1787, and four years later employed a force of seventeen, unusually large for that period. The founders were Colonel Russell White and Oliver Burr, a distant cousin of Aaron Burr. They turned out felt, beaver, korum (muskrat), and ladies' hats, as many as 2,300 dozen a year. Far from elegant, the early Danbury hats were rough, heavy and unwieldy, and sold for from six to ten dollars apiece. The hatmaker bought his skins in a bundle and by hand removed and sorted the fur. Unlike cotton or wool, felt was neither spun nor woven; instead, the hair or fur fibers were interlocked through the action of moisture, heat and pressure—a unique process. The know-how of men like Benedict and

CONNECTICUT MANUFACTORY

LOTTERY,

For raiſing the Sum of three Thouſand two Hundred Pounds.

The Managers being under oath, and having given bond for the faithful diſcharge of their truſt, preſent the Public with the following

S C H E M E.

1 Prize of	5,000 Dollars, is		5,000
1	2,500	-	2,500
1	1,500	-	1,500
5	1,000	-	5,000
10	500	-	5,000
15	200	-	3,000
50	100	-	5,000
100	50	-	5,000
300	25	-	7,500
325	15	-	4,875
500	10	-	5,000
4,400	8	-	35,200
1 laſt drawn Blank,	-		760

5,709 Prizes,	85,335
11,358 Blanks,	

17,067 Tickets at 5 Dollars each, is 85,335
Not two Blanks to a Prize.

Subject to a Deduction of 12 and an half per Cent.

This Lottery was granted by the honorable General Aſſembly for the encouragement of a Manufactory of Woolen, Worſted, and Cotton, in this State, under the ſuperintendance of William M'Intoſh, (late of London) a Gentleman of Information and Experience in the conſtruction and uſe of the new invented Machines for that Purpoſe, a Number of which being completed he hath now in uſe.

The Managers flatter themſelves that all Perſons will become Adventurers in this Lottery, who conſider the importance of the Object for which it was granted, as they will thereby aid one of the moſt valuable Manufactories attempted in this State, ſince the era of Independence.

They contemplate a ſpeedy ſale of the Tickets, and engage a punctual payment of the Prizes, if demanded in ſix Months after drawing, which is to commence on the 21ſt day of October next, and when finiſhed, the fortunate numbers will be publiſhed in the Connecticut Journal.

TIMOTHY JONES,
HENRY DAGGETT,
ELIAS BEERS, } Managers.
WILLIAM LYON,
NATHAN BEERS.

New-Haven, May 16, 1794.

Tickets to be had of the Managers, and of Thomas Hilldrup, at the Poſt-Office Hartford.

A lottery for a New Haven textile mill, 1794

Burr, combined with the abundance of water and fur locally, soon made Danbury the nation's leading hatting town.

In Waterbury, later the birthplace of the brass industry, wooden clocks and gilt buttons were being made. In New Haven, John Cook, the earliest carriage maker, turned out two-wheeled gigs.

At the eastern end of the state, in Mansfield, an enthusiastic group of farmers, spurred on by a state bounty, had become entranced with the exotic and demanding culture of the mulberry tree and the raising of silk. Commented Constant Southworth:

> . . . There is made in this town the present year (1791) about Two hundred-pounds weight of raw silk after being properly wound from the Cocoons and dried; on which there is a Bounty given by Government of two pence per Ounce . . . The business of winding, or as it is commonly called reeling the Silk, is now well understood . . .

One of the earliest charters granted by the legislature in 1789 was to the "Directors, Inspectors and Company of the Connecticut Society of Silk Manufacturers," but the act of incorporation gave no special advantage, and the enterprise suffered from the lack of skilled hands. Although Jeremiah Wadsworth wore silk ribbons from Mansfield to tie up his hair, his promotional efforts in Congress bore no more fruit than his backing of the Hartford Woolen Manufactory.

*   *   *

Despite the existence of embryonic manufacturing endeavors and their portent for the future, the maritime trade remained the most significant commercial activity, just as in Massachusetts and New York. It gave rise to the first real industry, shipbuilding, that boomed until 1820, and also to that class of able, daring, successful merchants who, when the heyday of shipping began to fade, lost no time in transferring their capital and their wits to manufacturing. Hartford, Middletown, New Haven, New London, Norwich and Stonington were bustling ports, where ships were built and berthed, the waterfronts chock-a-block with warehouses, riggers, ship chandlers and stores filled with tempting imports from the West Indies. With its 3,500-foot Long Wharf, New Haven led in size. New London was the leading shipbuilder, while Middletown became the busiest and richest riverport.

Connecticut's second largest industry, with an annual volume of half a million dollars, was the distillation of rum. Unlike shipbuilding, it was not specialized but rather an extra source of income for the farmer. There were over 500 one-man distilleries, the majority in Hartford County. Each still made only a few gallons of rum daily from the sugar and molasses brought back from the West Indies.

As the eighteenth century drew to a close, and for another five decades, the overwhelming majority of Connecticut Yankees depended upon agriculture for their livelihood. At the same time it was evident that the predominantly rocky soil could support no more. Only the rich loam along the banks of the Connecticut River could provide the basis for a thriving agro-business. Here in 1801 a Mrs. Prout of East Windsor made the first American cigars, known as "Long Nines." Farming methods were generally inefficient, but the most serious handicap was the farmers' lack of a broad market. True, New England did supply about half of the total exports of produce to the Caribbean, but very little to domestic markets. The only answer was emigration. Motivated by the need for new soil and by the lure of a life free of Puritannical restraints, many families moved to New Hampshire and Vermont, or out to the free lands of the western frontier in Pennsylvania and Ohio. In 1819, Pease & Niles wrote:

> The spirit of emigration which has prevailed so extensively in this State, disclosed itself previously to the Revolutionary War . . . Within the last thirty years the current of emigration . . . has swelled to a torrent, and has been directed principally to the westward.

Between 1790 and 1820 the population of New England grew at a mere one-third the national rate. Connecticut's increased from 238,000 to 275,000, or only 15.5 percent. Some towns—Colchester, Derby, Lebanon, Simsbury, Wallingford, Watertown, Windham and Woodbury—even suffered a decline. People were widely dispersed among 117 towns, only six of which in 1800 had more than 5,000 inhabitants: Hartford, Middletown, New Haven, New London, Stonington and Norwalk (which then included New Canaan and Wilton). New London and Stonington shrank considerably as the result of the maritime depression. By emigration the state probably lost during these years almost as many as it contained in 1790. Most of this loss flowed from the noncommercial inland towns. On the other hand, Danbury with its hatters' shops gained twenty percent, from 3,031 to 3,606. Not until manufacturing came into its own did the drain of Connecticut natives stop and the influx of new settlers from Europe commence.

At the dawn of the nineteenth century, therefore, farming and its close ally, homespun industry, ruled the economy. There were no manufacturing towns. Rather than being a specialized producer, the artisan was simply a farmer who in idle seasons supplemented his income by using special skills acquired out of the necessity to be self-sufficient. His ability as a Jack-of-all-trades was homespun's greatest contribution to the future of industry.

Thus, a combination of forces had already begun to turn the Yankee irrevocably toward a new calling: manufacturing. These included the unavailability of more tillable soil to support additional people; no national market for surplus food products; the opposition of English merchants to the development of domestic manufacturing; the natural inventiveness and restless energy of Anglo-Saxon Congregationalists; the choice between self-sufficiency and starvation; and the spiritual injunction to seek a personal occupation worthy of the individual.

Another contributing factor was the evolution of a distinctive democratic tradition, beginning with Thomas Hooker's Fundamental Orders. From his concept of government, which made individual duty and right living the supreme virtues, came more political leaders, educators, entrepreneurs and shrewd traders than any other colony produced. It led to that unique institution of local self-rule—the town meeting, which is still retained by over 100 Connecticut towns.

Population of Leading Towns in Connecticut

	1790	1800	1810	1820	1830	1840	1850	1860
Bridgeport					2,800	4,570	7,560	13,299
Bristol	2,462	2,723	1,428	1,362	1,707	2,109	2,884	3,436
Danbury	3,031	3,180	3,606	3,873	4,311	4,504	5,964	7,234
Hartford	4,090	5,347	6,003	6,901	9,789	12,793	13,555	29,152
Middle-town	5,375	5,001	5,382	6,479	6,892	7,210	8,441	8,620
New Britain							3,029	5,212
New Haven	4,484	5,157	6,967	8,327	10,678	14,390	20,345	39,267
New London		5,150	3,238	3,330	4,356	5,519	8,991	10,115
Norwalk	11,942*	5,146	2,983	3,004	3,792	3,863	4,651	7,582
Norwich		3,476	3,528	3,634	5,179	7,239	10,265	14,048
Stamford		4,352	4,440	3,284	3,707	3,516	5,000	7,185
Stonington		5,437	3,043	3,056	3,401	3,898	5,431	5,827
Waterbury	2,937	3,256	2,874	2,882	3,070	3,668	5,137	10,004
Wethers-field	3,806	3,992	3,961	3,825	3,853	3,824	2,523	2,705
State Total	237,946	250,902	261,942	275,248	297,675	310,015	370,792	460,147

* Including Greenwich and Stamford.

SOURCE: Connecticut State Register & Manual, 1969, pp. 578–82.

Concurrently, things were happening in transportation, finance and education that would spur industry on. The first turnpike company, from New London to Norwich, was incorporated. Banks were opened in Hartford, New Haven and New London. A fire insurance company appeared in Norwich, a marine underwriter in New Haven. The sale of its Western Reserve lands in Northeastern Ohio for the sum of $1,200,000 enabled the state to set up a School Fund that greatly furthered the growth of publicly-supported elementary schools. Connecticut, along with its sister colonies Massachusetts and New York, stood on the threshold of a social and technological revolution unparalleled in American history until the industrial expansion after the Civil War or the arrival of the Space Age after World War II.

Succeeding chapters will show how Connecticut Yankees pursued their calling, as manufacturers and itinerant peddlers, as founders of factory villages and the American factory system, in the evolution of corporations, in mass producing clocks and various wares, in the rise and decline of entire industries and in the proliferation of inventions—all of which culminated in the predominance of manufacturing by the time of the Civil War.

CHAPTER II

The Factory Village

TRAVELLING through the older New England towns with their concentration of gray mills, brick factories and towering smokestacks might lead one to conclude that industry began in the city. Actually, manufacturing took place wherever there was an adequate source of water for power, but especially in rural areas where entirely new communities were spawned. Only after the middle of the 19th century did it concentrate in the cities in order to take advantage of steam power for running machines, of the railroad for distribution of its products and of a more adequate supply of labor. Today, in a reverse trend, the factory has again returned to what is now the suburban countryside to satisfy its need for space.

The manufacturing village was in many ways a unique creation of the early Yankee entrepreneur. It was one of the first planned communities, in a few instances aspiring to be a kind of economic utopia, with company-built boarding houses, schools, libraries, churches and company-operated stores. Some employers assumed responsibility for their workers' moral character and well-being. Child labor, in contrast to the neighboring state of Rhode Island, was minimal. The prevalence of these villages between 1800 and 1850 and the variety of their products can best be appreciated by reading the summary included at the end of this chapter.

Within the five thousand square miles of Connecticut are 169 townships. In addition, there was at one time the astounding total of 203 "villes", mostly independent units formed for a particular economic purpose and adjacent to some river or stream. Most were founded for manufacturing purposes; many were named for the founder himself. Thirty-three are still viable enough to have a post-office; sixty-six have disappeared entirely; only two matured into townships. Formerly part of New London, Montville, the earliest, became a town in 1786. Here the paper and textile industry got a hold. Plainville, named after the "Great Plain" between Farmington and New Haven, came into being in 1831 as boats on the Farmington Canal were carrying sugar, salt and other products upcountry from the port of New Haven. Plainville's expectations of becoming an important industrial center never materialized because of the canal's failure in 1847.

Those "villes" no longer in existence have left behind some quaint and descriptive names: Fluteville, near Litchfield, known for wooden flutes in the 1830s; Hitchcocks-ville, home of the famous chairs, renamed Riverton in 1866 to end the postal confusion with Hotchkissville; Hoadleyville, site of Seth Thomas's first clock factory; Johnsonville on the Moodus River in East Haddam where cotton twine was spooled; Spoonville in East Granby, silver plating. Smaller villages included Almyville (cotton yarn); Augerville (augers); Burtville, the original home of Borden's condensed milk; Dobsonville (cotton); Granite-ville in Waterford; Greenmanville (shipbuilding) in Stonington; Griswoldville (edge tools and hammers); Gurleyville (silk); Shailerville (gristmill); and Whig-ville, once the largest copper mine in the state.

Among those still maintaining a postal identity are: Ballouville (cotton yarn) and Dayville (cotton) on the Five Mile River; Forestville (wooden clocks); Mechanicsville (textiles); Oakville (pins and wire goods); Plantsville, named after the Plant family who manufactured buttons and tinware, near Southington; Uncasville (cotton); Unionville (wood screws and spoons); Westville (matches); and Yalesville, home of the Yale family's Britannia ware.

It is interesting to speculate why Connecticut, as well as Massachusetts, Rhode Island and Pennsylvania, used the suffix "ville" for place name endings in such profusion. Why not instead Quarrytown or Plainvillage? England had no "villes" that could set a precedent. Being, of course, the Gallic name for village, "ville" suggests French influence. Five years before the naming of Montville, 6,000 French soldiers under Comte de Rochambeau marched through there on their way to Yorktown. Yet some hardheaded Anglo-Saxon Yankees did not take kindly to French frills, even for place names. When Collinsville was so named in 1826, the axe-maker Samuel Collins vigorously objected; he would have preferred "Collinsford" or "Valley Forge".

The use of "ville" was almost entirely post-Revolution, replacing the English "borough" and Scottish "burgh". An authority on place names points out that "enthusiasm for things French ran high in those years . . . in a very short time -ville became so well Americanized that few people thought of its ever having been anything else . . . No feature of American naming has provoked fiercer attack than the prevalence of this suffix." After 1850, however, founders of villages discarded suffixes, and "villes" began to pass into antiquity as the city emerged.

Visitors to the factory villages found them generally drab and grimy. The row-upon-row of white frame houses; the noisy, dusty mill always close to a source of water power; the frequent absence of graceful church spires or one-room schoolhouse—all these were in sad contrast to the charming greens of the older towns, many of which were left behind in time and space as the new mill sites became the center of economic activity. Yet, such unpleasant conditions were by no means universal. Depending upon the personalities and aims of the founders, the appearance, layout and operation of the factory villages varied considerably. Some evidenced considerable charm. Many contained the amenities essential to a wholesome and uplifting life. All had a cohesiveness not found in larger towns. In most cases, living conditions were superior to those prevalent in English mill towns, as well as to those in the congested slums that accompanied the later growth of factory cities like Hartford, New Haven and Bridgeport.

From the beginning, textile mills, both in England and America, tended to represent the worst side of laissez faire philosophy. In theory it offered freedom to both capital and labor, but in practice fostered human exploitation. Besides doing away with the domestic artisan, English mills took children from the almshouses, paid pitifully low wages and were known for harsh treatment. Workers were crammed into miserable abodes that lacked any semblance of sanitation.

When Samuel Slater left England in 1789 at the age of twenty-one with the ambition of founding his own mechanized cotton factory somewhere in America, he brought with him a mixture of his country's distasteful labor practices and a personal concern for the worker's welfare. A chance encounter with a packet captain led him to Providence and to Moses Brown, a retired Quaker merchant of considerable means. Brown, having tried without success to produce cotton yarns on crude, hand-operated equipment, desperately needed an experienced overseer to save his investment. Without delay he hired Slater with the inducement that he would "have the *credit* as well as the advantage of perfecting the first watermill in America". Within a year, in Brown's Pawtucket mill, the young English mechanic lived up to his promise of making the first cotton yarn by machine in this country.

As soon as he had reproduced Arkwright's spinning frame, Slater also introduced the *social* institutions promoted by English employers. For his hands he built the first Sunday school in New England, offering instruction in the three R's as well as in religion. Later he encouraged common day schools, in some cases paying the teachers out of his own pocket. Paternalistic to a degree, Slater was well aware of the evils that already bedeviled English mill towns and exerted his best efforts to avoid their recurrence in his adopted homeland.

In 1806, the textile firm of Almy, Brown & Slater erected another mill with Slater's brother, John, in charge. "This event," claimed Edward Stanwood, "may be taken as the first example, in this country, of the creation of a factory village,

Samuel Slater

The mill at Slatersville

Whitney's cotton gin

out of which have grown the factory town and the factory city." On the invitation of his elder brother, John Slater had emigrated three years before. He was soon exploring the countryside on horseback for a suitable site and selected Smithfield on the Blackstone River because of its forty-foot waterfall. Previously, new mills had been erected only where a settlement existed large enough to supply local labor. Since Smithfield was then a wilderness with no advantages except its ample water supply, the Slaters had to form a community from scratch. By 1810, a two and one-half story schoolhouse and mill tenements for four families each, as well as three mills, had been raised. The name was changed to Slatersville, the cotton operations becoming the largest and most advanced in Rhode Island with 300 looms, 12,000 spindles and 320 employees.

Although John Slater undoubtedly shared his brother's enlightened philosophy, the Slaters' good intentions failed to offset the bad effects of English labor practices. The basis of the Rhode Island system, as it became known, was the employment of whole families, especially children, who were considered more agile and dexterous than their parents. In Samuel Slater's first mill nine children between the ages of seven and eleven comprised most of his small working force. Unfortunately, the majority of Rhode Island employers in copying Slater's production ideas disregarded the humane aspects of his system. To them workers were no different from machines—to be worn out and replaced as necessary. Unlike him, they made almost no effort to provide decent living accommodations. Minors worked the same twelve-hour day as adults, but they were apparently given several play periods during the day in consideration of their age. Wages were perhaps a little higher than on the farm, but still very low (forty-two cents a week) and paid only once a year. Furthermore, they had to be taken out in trade at the company store. More so than any other state and for a longer period, Rhode Island perpetuated the European tradition of child labor. By 1831 it employed nearly 3,500 under the age of 12, or three-quarters of all those in the nation's cotton mills. On the other hand, Connecticut had only 439 and Massachusetts none.

Happily, the Rhode Island labor system did not take root in the United States, even though cotton manufacturers in New York and New Jersey followed the Slaters' example. If it had become the sole pattern for the social organization of workers in America, it is conceivable that, like their English brethren, they would have been more class-oriented and more trade-union-minded at a much earlier date. In contrast, the Waltham system, introduced by Francis Cabot Lowell and a group of wealthy Boston merchants in 1813, was more sensitive to the moral and reform traditions of their Puritan ancestors. They determined not to recreate the notoriously degraded conditions in and around the mills of Birmingham and Manchester, England. Their objective, instead, was to attract the best workers available under carefully controlled working and living conditions.

The nucleus of the Waltham system was the company boarding house or dormitory. Young intelligent girls from good farming families were hired and dressed in neat white gowns. Everything was done to create a respectable environment that would "overcome the prejudice of conservative rural New England against industrial employment." Wages were paid monthly and in cash. Even so, the lot of the "nuns of Lowell" was far from ideal. Just as on the farm, they worked from sunrise to sunset, although probably at a more leisurely pace than would be tolerated in most factories today. Mills were poorly ventilated and heated, dormitories overcrowded and unsanitary. In 1839, in fact, Lowell came under sharp attack from the Boston press, and the virtues and drawbacks of the whole system in Massachusetts were hotly debated by the newspapers and the legislature. By mid-century the girls were driven out by the failure to uphold the initial employment standards. A host of immigrants, mostly Irish, replaced them, ending the attempt to make Lowell a puritanical, paternalistic paradise and ushering in the era of the urban mill.

Defenders of the factory village were as vociferous as its critics. In his *Memoir of Samuel Slater*, published in 1836, George S. White devoted an entire chapter to the "Moral Influence of Manufacturing Establishments," which he said:

> became a blessing or a curse according to the facilities which they create for acquiring a living, to the necessary articles which they provide, and the general character which they produce. To set up and encourage the manufacturing of such articles, the use and demand of which produces no immoral tendency, is one of the best and most moral uses which can be made of capital. The moral manufacturer . . . is in reality a benefactor.

To bolster his defense of the factory village, White had circulated a questionnaire to several employers in New York and New England. One respondent was Smith Wilkinson, whose family had established a cotton mill on a thousand acres of wilderness around Pomfret, Connecticut, on the Quinebaug River:

> The usual working hours, being twelve, exclusive of meals, six days in the week,—the workmen and children being thus employed, have no time to spend in idleness or vicious amusements. In our village there is not a public-house or grog-shop, nor is gaming allowed in any private house . . . In collecting our help, we are obliged to employ poor families, and generally those having the greatest number of children . . . These families are often very ignorant, and too often vicious; but being brought together into a compact village, often into the families and placed under the restraining influence of example, must conform to the habits

and customs of their neighbors . . . I have known a great many, who have laid aside $200 to $300, in from three to four years . . . Perhaps I cannot furnish better proof of the practical tendency and effect on female character, than to state, that in 29 years, during which term I have had the sole agency of Pomfret cotton manufacturing establishment, I can assert that but two cases of seduction and bastardy have occurred.

Besides Smith, the youngest son, four other Wilkinson brothers (including David, the inventor of the lathe) and two sons-in-law were involved in the company as stockholders or employees. When the mill roof was raised on July 4, 1806, 2,000 people assembled to celebrate with free punch. A capable manager, Smith ran the factory village he created with a stern hand. The main purpose of buying so much land, he asserted, was not only to prevent the sale of spirits, but to establish schools and to introduce public worship on the Sabbath:

> Accounts and morals were looked after with equally keen scrutiny. No man was allowed to overrun his credit, get drunk or misbehave on Sunday. Religious services were held in the brick school-house whenever practicable . . . Pomfret Factory was remarkably exempt from the immoralities and disorders incident to the ordinary factory village of the period. Houses and yards were kept neat, loafing prohibited, children sent to school as the law required . . . Yet notwithstanding this severity of discipline, there was much that was pleasant and enjoyable in life at the old Pomfret Factory.

Another manufacturer from Troy, New York, told White that there was greater attention paid to schooling children in manufacturing villages than anywhere else. The letters received by White mentioned, too, the customary practice of providing boarding houses, the healthiness of steady labor, the advantage of the company store to neighboring farmers as well as mechanics and the improvement in education and character that resulted from living in such wholesome surroundings. If one accepted the White thesis at face value, the manufacturing village was responsible for reclaiming, civilizing and Christianizing hundreds of illiterate, disorderly and irreligious poor families throughout New England.

The American Home Missionary Society, however, painted a much less idyllic picture in 1848. Aware of the almost life-and-death power of the capitalist over the hapless wage earner, the missionaries feared that manufacturing villages were altogether too barren of such amenities of good living as churches, schools and comfortable dwellings. The Society took the position that the company had the same obligations to its workers as the master did for his apprentice: "Corporations are morally bound to exercise a beneficent parental care over the social and moral welfare of the villages which they create and control." In addition to churches and schools, corporations were urged to provide libraries, social halls, lectures and pleasant surroundings. Both mills and villages should be kept neat and clean, since there was "always some immorality in dirt." Above all, the companies "should keep with the utmost strictness the holy Sabbath of God." The Society condemned the twelve-hour day as leaving too little time for intellectual improvement, social enjoyment or religious culture, and it deplored the employment of idle, thriftless and dissolute operatives who were prone to drunken sprees.

For example, five years after the Wilkinsons started their model community, another cotton mill arose nearby in a swampy hollow on the French River that earned a quite different reputation. Founded by the Masons and others, the Thompson Manufacturing Company quickly attracted laboring men and their families, young men and girls, transforming the lonely valley into a brisk little village, known for many years as "The Swamp." A Baptist preacher, Elder Pearson Crosby, condemned "Swamp Factory" as a place "where for two or three years Satan had seemed to reign with almost sovereign and despotic sway. Vice and immorality (were) permitted to riot without control. The sound of the violin, attended with dancing, the sure prelude to greater scenes of revelry for the night."

The Missionary Society issued this final warning: "Manufacturers will find it to their interest to pursue a liberal policy in their villages, in order to check the spread of that opposition to corporations which . . . if manufacturing villages become the abodes of a low and degraded population, will sweep everything before it, in the indignation of the people."

* * *

Although Slater must be credited with being the first employer in New England to show concern for the well-being of his laborers, the village bearing his name was not the first manufacturing community in America. That honor, it seems clear, must be accorded Whitneyville in Connecticut.

Eli Whitney had already seen his cotton gin revolutionize Southern agriculture. He had gained from his invention, however, nothing more than frustration, debt and impaired health. In the spring of 1798 Whitney indomitably turned right-about-face to enter into the mass production of guns for the U.S. government. He wrote his good friend Oliver Wolcott Jr. in Philadelphia, then Secretary of the Treasury:

> I am persuaded that machinery moved by water . . . would greatly diminish the labor and facilitate the Manufacture of this Article . . . There is a good fall of water in the vicinity . . . which I can procure, and could have works erected in a short time.

Actually, when he signed a contract for 10,000 muskets, the 33-year-old inventor had neither experience nor a factory site, workmen, tools nor machines. In September, he bought 100 acres along the Mill River, where New Haven's settlers had put their first gristmill. An old, six-foot log dam provided a year-round waterfall. Timothy Dwight, then president of Yale College, wrote that "no position for a manufactory could be better. From the bleak winds of winter it is completely sheltered by the surrounding hills . . . . No place, perhaps, is more healthy; few are more romantic."

Whitney encountered difficulties in completing the purchase. Not until November 1 could he rebuild the dam and erect buildings, which he rushed to occupy before the onset of what proved to be a long and severe winter. On January 13, 1799, he could exult that "my building which is 72 feet long by 30 & 2 stories high is nearly finished!" Soon he added shops for stocking wood, metals, leather and glass, and for a trip-hammer, each served by a flume; across the river at the foot of Mill Rock, in order to reduce the fire hazard, he located the forging shop, with seven pairs of bellows and a cluster of storehouses.

Quickly he recruited, mostly from Massachusetts, some fifty skilled hands. For those with families he built five handsome stone houses, and for the unmarried he ran a boarding house. A bachelor himself, he lived in a farmhouse opposite the mill across the Hartford turnpike, along with three nephews, a dozen apprentices and servants. A stone store served the little community, which—like Slater's—was paternalistic. However, lacking the Englishman's religious zeal, Whitney did not insist on his employees attending church. Tireless, demanding, seething with ideas and plans, he was bothered, not by morals, but by stupidity and ignorance. Although his consuming passion for work, work, work to the complete exclusion of the outside world left him no time for worrying about such small details as churches or schools, he did show an interest in training young men as mechanical apprentices. By 1803, his factory had become a landmark.

* * *

The same year that Slatersville sprang up, some 125 miles to the west, Salisbury ironmakers were finishing atop Mt. Riga the largest furnace yet seen in that area, where a rich vein of iron ore had been mined and refined for seventy-five years. Around it, in 1810, grew a village or, more accurately, a workers' camp, that had all the attributes of future frontier boom towns. As many as fifty furnaces, their forges and charcoal pits illuminating the countryside at night like giant torches, were then scattered about Litchfield County. Three stories high, they consumed 250 bushels of charcoal and three tons of the native ore to produce one of pig iron. Importing fifty or so colliers, supposedly from as

Whitneyville, 1826–28

far away as Switzerland and Lithuania, Holley & Coffing finished its big new furnace in time to supply cannon, rifle barrels and anchors for the War of 1812. This partnership became the leading ironmaker of Salisbury for twenty-five years, also owning works at Lakeville—called "Furnace Village"—and Lime Rock.

Three miles north of, and 1,300 feet above, Salisbury, Holley and Coffing built a spacious home for their ironmaster, Joseph Pettee, and accommodations for the workmen, few of whom had families. Yet within a decade they ran a school for seventy to eighty pupils and a store that kept four clerks on the jump and stocked the best imported silks. With an annual payroll of $150,000, Mt. Riga soon grew into the gayest, wealthiest village in the district. Hard drinkers and avid hunters, Riga men were known as "the Raggies". Their origin and even that of the name Mt. Riga are both shrouded in mystery.

The chief advantages of Mt. Riga were its unlimited supply of wood for charcoal and abundance of water from Forge Pond near the summit, although iron ore and limestone had to be laboriously carried up the mountain three or

Forbes blast furnace, East Canaan

Holley Manufacturing Company, Lakeville, 1844. Oldest pocket knife factory in America

four miles by packhorse. Wagonloads of charcoal were stored in sheds. The works extended a half mile along a stream; half of the pig iron was processed in two forges, one for anchors and plows, the other for chains, tools, hinges and utensils. The most notable forgings were anchors for U.S. ships like the *Constellation* and *Constitution*. Pettee would invite the appropriate Navy officers to an inspection ceremony, where the anchors were hoisted into the air on tripods and then dropped hard to the ground. If they survived this test, they were loaded onto oxcarts and dragged downhill all the way to the Hudson River for shipment. Afterwards there was a gay ball attended by the officers and plenty of pretty girls.

For two decades Salisbury furnaces supplied most of the musket barrels for the government arsenals at Harpers Ferry and Springfield. Until the late forties, 5,000 or 6,000 tons of the brown hematite, yielding about forty percent pig iron, were dug each year and sold for three dollars a ton. Noted for its superior tensile strength, Salisbury iron was favored by such meticulous manufacturers as Eli Whitney for his muskets and Samuel Collins for his axes.

The Mt. Riga Furnace, however, lasted only until 1847, by which time the railroad had opened up cheaper sources of ore. As for the displaced Raggies, they

General David Humphreys

and their descendants degenerated into a lost, hopeless band of mountaineers who failed to adapt to other employment. Many ended up living in squalid shacks along the lower slopes of the mountain. The abandoned homes of Mt. Riga, built on stilts without cellars, quickly fell into ruins.

*   *   *

For nearly a half century, until 1850, Rimmon Falls in Derby, some four miles west of the Naugatuck River, was the site of the earliest planned factory village—Humphreysville, named after one of Connecticut's most versatile, popular and patriotic pioneers. General David Humphreys, son of a Derby minister, graduated from Yale College in 1771 at nineteen, joined the Continental Army, became Washington's personal aide, and had the honor of presenting to Congress the British flag surrendered by Cornwallis at Yorktown. From soldier he turned poet, becoming a Hartford Wit along with John Trumbull, Joel Barlow, Lemuel Hopkins and Timothy Dwight. With the latter he co-authored the *Anarchiad*. As Federalists, the Wits "were the literary old-guard of the expiring eighteenth century, suspicious of all innovation . . ." Not so Humphreys. Under his huge, portly frame and pink complexion, he harbored thoughts that were for the times controversial and in the long run antithetical to Federalist doctrine. He was the electrical charge that closed the gap between the philosophies of Jefferson and Hamilton—between farm and factory.

From poet he turned ambassador in 1791, first to Portugal and then to Spain. In Lisbon his marriage to the very wealthy daughter of an English banker, Ann Frances Bulkley, enabled him to pursue vigorously for the rest of his life his ideas on agricultural reform and the development of domestic wool manufacturing in the United States. A few years before, in the last of his patriotic poems, he had lyrically expressed his dream:

> First let the loom each lib'ral thought engage,
> Its labours growing with the growing age;
> Then true utility with taste allied,
> Shall make our homespun garbs our nation's pride.
> See *wool*, the boast of Britain's proudest hour,
> Is still the basis of her wealth and pow'r!
> Then, rous'd from lethargies-up men! increase,
> In every vale, on every hill, the fleece!
> to toil encourag'd, free from tythe and tax,
> Ye farmers sow your fields with hemp and flax:
> Let these the distaff for the web supply,
> Spin on the spool, or with the shuttle fly.

In Spain, observing the hardiness of the Merino sheep, he charmed the Spanish grandees, who jealously guarded their monopoly of the fine wool market, into allowing him to take home as a parting gift twenty-five rams and seventy-five ewes. They were driven across country to Lisbon, thence shipped to New York in April, 1802. All but nine survived the ocean passage; upon reaching New York, they were transferred to a sloop bound for Derby, where they created intense excitement. This flock became the source of most of the early pure-blooded sheep in America, although for five years farmers confined their interest to mere curiosity. Humphrey's animals thrived. Their wool being as fine as that grown in Spain, was valued at $1.00 per pound, double that of domestic wool. Passionately convinced of the timeliness of his experiment, he wrote a Boston friend:

> The importance of an internal supply of the first articles of necessity appears to be more understood and acknowledged every day . . . It may be asked, How long are we to continue thus like colonies dependent on a mother country? And will a period never arrive when it will be indispensable to clothe ourselves principally, with our own productions and fabrics?

> The period of a general peace promises more than any other to promote and accelerate the establishment of manufactures. That event, by producing a stagnation in our foreign navigation, nearly closing the avenues to commercial speculations and diminishing the external demands for our provisions, will afford a favourable opportunity to invest a part of the surplus capitals in this manner.

In 1807, America was shut off from woolen imports by the Embargo, patriotism stirred the American heart, and Humphrey's hopes for wool growing and the wearing of homespun suddenly came to life. Merino sheep importing and raising quickly became a frenzied speculation, much like the silk fever of the 1830s. Wool production soared. By the time Congress increased ad valorem duties on raw wool to thirty-five percent in 1812, Humphreys was able to sell a ram for as high as $2,000 and ewes for $1,600, compared with the price of $100 which he charged Derby farmers a decade earlier.

Hardly had his Merino sheep set their hooves on Derby soil than General Humphreys determined to establish a woolen factory in his birthplace. For the sum of $2,647.92, he purchased property around Rimmon Falls that included a sawmill, two fulling mills, a clothier's shop and a dam. Together with Captain Thomas Vose of Derby he formed a partnership, journeyed to England to study factory methods, and brought back some weaving equipment along with an experienced woolen man named John Winterbotham. He lured other skilled workmen from abroad and, in 1806, the same year Slatersville and Pomfret came into being, erected a two-story woolen mill. The next year Humphreys acquired almost 300 acres up to the Naugatuck River. His industrial experiment started out as successfully as his breeding of sheep. Displays of his cloth at local agricultural fairs won acclaim. In 1810, he obtained a charter to incorporate the Humphreysville Manufacturing Company with a capitalization of $500,000.

From the beginning he intended to make Humphreysville a model village, a showplace to prove that domestic cloth could be woven profitably, that its quality could equal any import, that manufacturing could provide a new source of livelihood for the inhabitants, that working conditions (unlike those in England's "dark Satanic mills", as the poet William Blake described them) could be healthy and morally uplifting. Waterpower operated the wool-carding machines, jennies and looms. Owing to the depression in agriculture and shipping, labor was plentiful. Women and children did most of the work, especially hand weaving, earning up to four dollars monthly. Skilled men received as much as twenty-one dollars a month. Humphreys strove to produce goods of the best quality, knowing they would be the most saleable. He built homes for his workers.

At first parents were reluctant to place their children in the mill, forcing Humphreys to turn to pauper apprentices, as the English cotton manufacturers had done earlier. From the New York almshouse he obtained seventy-three boys. He took great interest in the discipline and education of these orphans, providing Sunday and evening schools in the manner of Samuel Slater. He was a stickler for good manners and proper behavior. Once assured of his moral rectitude, his neighbors offered their sons and daughters in more than sufficient numbers.

His enterprise, which he looked upon as more philanthropic than profit-making in purpose, soon achieved national prominence. Humphreys's efforts helped to change the attitude of the leading opponent of manufacturing, none other than Thomas Jefferson. In November, 1808, the President wrote a New Haven friend:

> Homespun is become the spirit of the times. I think it an useful one, and therefore that it is a duty to encourage it by example.—The best fine cloth made in the United States I am told, is at the manufactory of Col. Humphreys . . . Could I get the favor of you to procure me there as much of his best as would make a coat?

He got his coat in time to wear to a New Year's Day exhibition at the White House, at which he anticipated every one would be dressed in American-made cloth. The Colonel rushed him five and one-half yards, for which Jefferson paid $24.75. In his letter of thanks Mr. Jefferson told Humphreys:

> (The cloth) came in good time & does honour to your manufactory, being as good as any one would wish to wear in any country. Amidst the pressure of evils, with which the belligerent edicts have afflicted us, some permanent good will arise, the spring given to manufactures will have durable effects . . . My idea is that we should encourage home manufactures to the extent of our own consumption of

Humphreysville

Hotchkiss & Merriman's gum elastic manufactory, Waterbury

everything of which we raise the raw material. I do not think it fair in the ship-owners to say we ought to make your own axes, nails &c. here . . .

With which, of course, the Federalist Humphreys heartily concurred, although he must have been amazed at the President's retreat from his former hostility to domestic industry. Jefferson, however, never went so far as to embrace the factory system, which Humphreys was in effect fostering.

The Connecticut legislature in 1808 praised him for his "patriotic exertions" in importing the first Merino sheep into the state and for his "wise and well-considered measures . . . in establishing and conducting the manufacture of cotton and woollen fabricks at Humphreysville . . ." In appreciation and to encourage him further, the legislators, with Governor Jonathan Trumbull's approval, exempted his workers from the poll tax, military duty and working on state highways. They also gave his mills a ten-year moratorium on paying taxes. This action served to popularize his village even more and to accelerate the demand for Connecticut-made woolens.

The best description of Humphreysville was recorded by Humphreys's close friend, the verbose chronicler, Timothy Dwight:

> Within the limits of Derby, four miles and a half from the mouth of the Naugatuc, is a settlement named by the Legislature Humphreysville, from the Hon. David Humphreys, formerly Minister plenipotentiary at the Court of Madrid. At this place a ridge of rocks, twenty feet in height crosses the river, and forms a perfect dam about two thirds of the distance. The remaining third is closed by an artificial dam. The stream is so large, as to furnish an abundance of water . . .
>
> A strong current of water in a channel, cut through the rock on the Eastern side, sets in motion all the machinery, employed in these buildings. By this current are moved the grist-mill; two newly invented shearing machines; a breaker and finisher for carding sheep's wool; a machine for making ravellings; two jennies for spinning sheep's wool, under the roof of the grist-mill; the works in the paper-mill; a picker; two more carding machines for sheep's wool; and a billy with forty spindles in a third building; a fulling-mill; a saw-mill, employed to cut the square timber, boards, laths, &c. for the different edifices, and to shape many of the wooden materials for the machinery; two more fulling mills on improved principles, immediately connected with the clothier's shop; and the various machinery in a cotton manufactory, a building about one hundred feet long, thirty-six wide, and of four stories, capable of containing two thousand spindles with all their necessary apparatus.
>
> The houses can accommodate with a comfortable residence about one hundred and fifty persons. Ten others in the neighbourhood will furnish comfortable residences for upwards of one hundred and fifty more. Gardens on a beautiful plot

in the rear of the manufactories, furnish all the vegetables, necessary for the establishment.

The institution contains four broad and eight narrow looms, and eighteen stocking-frames.

The principal part of the labour in attending the machinery, in the cotton and woollen manufactories, is done by women and children; the former hired at from fifty cents to one dollar per week; the latter, apprentices, who are regularly instructed in reading, writing, and arithmetic.

The wages of the men are from five to twenty-one dollars per month.

In Europe great complaints have been made of manufacturing establishments, as having been very commonly seats of vice, and disease. General Humphreys began this, with a determination either to prevent these evils, or if this could not be done, to give up the design. With regard to the health of his people it is sufficient to observe, that from the year 1804 to the year 1810, not an individual, belonging to the institution, died . . .

With respect to vice it may be remarked, that every person, who is discovered to be openly immoral, is discharged.

The manufactures at Humphreysville are esteemed excellent. The best broadcloth made here, is considered as inferiour to none which is imported.

To placate those who regarded factories as workhouses of sin and who wanted to restrict manufactures to the home, Humphreys's charter of incorporation in 1810 contained this unique provision:

It shall be the duty of the president and directors . . . to provide an instructor, for at least three months in each year, for the purpose of teaching the children employed . . . to read and write, and also the first four rules of arithmetic, and in religion, morals and manners, as is by law directed to be taught in other schools.

Three years later, while serving as a state representative himself, he persuaded the lawmakers to amend the old master-servant relationship by extending this provision to all other manufacturing companies. Entitled "An Act relating to masters and servants, and apprentices," it was the first piece of general labor legislation passed in Connecticut but never diligently enforced. It required the selectmen in each town, as a board of visitors, to inspect every factory annually to make sure child workers were properly educated and their morals protected. If a mill was found in neglect of these duties, the county court could abrogate the indentures or contracts binding the children or fine the partners or owners up to $100.

Common schooling had existed in Connecticut since 1650. Loosely organized into over 200 societies and districts, and administered by the local citizenry, schools were, in Humphreys's day, overcrowded and understaffed. Each school was allowed but one underpaid teacher, who took care of fifty or more pupils. School attendance was not obligatory, and one in five children did not attend at all. Not until a state board of commissioners was appointed in 1838 to investigate school conditions and Henry Barnard, the Hartford educator, was hired as secretary, did reforms begin.

Some years afterward Humphreys suggested to James Monroe, then Secretary of State, that his works offered "one of the most eligible sites in the U.S. to found an Institution for extending the benefits of a Military & Manufacturing Education to the Orphan or other poor children of Soldiers or other Citizens, at the least possible expense to the Public . . . If in any thing, I have had an opportunity of being useful to my Country . . . more so by setting an example of educating youths at the Humphreysville Establishment, than in any other way." Education, whether of farmers or orphans, was clearly his foremost love.

The end of the War of 1812 left New England's infant manufacturing temporarily depressed. Struggling to survive besides Humphreysville were twenty-eight mills employing 1,200 persons. They produced about 100,000 yards of cloth a year, only one-fifth of that spun in households. Humphreys strongly felt the need for protection of domestic manufacturing from foreign competition. In 1816 Congress, through the efforts of Henry Clay, did pass the first protective tariff on woolen goods. In a stirring address on husbandry that same year, Humphreys urged Connecticut farmers to stop emigration by enriching their lands through hard work, the spreading of manure and the rotation of crops. Pointing out that the day of the merchant and shipowner had been eclipsed, while manufactures languished, he argued that only perseverance in agricultural improvements could restore commerce and industry. The following year, under his guidance, the Connecticut Agricultural Society was incorporated, and he became its first president. Thus, at the end of his life two years later, David Humphreys evinced more faith in agricultural reform than in manufacturing. His imaginative venture at Humphreysville, however, earned him the posthumous reputation of having been "the founder of the New England factory village," according to the National Association of Wool Manufacturers in 1899. In retrospect, this giant of a man, this Yankee philanthropist, symbolized the transition taking place between the agricultural society of the 18th century and the rising industrialism of the 19th century.

* * *

Besides Jeremiah Wadsworth, Eli Whitney and David Humphreys Connecticut produced another manufacturing pioneer who achieved national prominence. Oliver Wolcott Jr., the third generation of his family to serve as governor, devoted most of his life to public service. For eleven years he held high

positions in the U.S. Treasury, succeeding Alexander Hamilton as the second
Secretary of the Treasury in 1795. After a short stint as a federal judge, as head of
his own trading company in New York and as a bank president, he returned to
his Litchfield farm. So many changes had occurred during his 26-year absence that
he felt almost a stranger in the land of his fathers. He quickly discovered that
Connecticut, as the result of the War of 1812 with its disruption of commerce,
was becoming a manufacturing state.

Wolcott's interest in manufacturing undoubtedly was stimulated by
Hamilton's vision of an industrial future for America that would enhance, rather
than hurt, agriculture and make the country self-sufficient. But through his long
friendship with Eli Whitney it became a cause he would pursue the rest of his
life. The two met when the young inventor showed him a model of his cotton
gin, and Oliver became his staunch supporter. When Whitney came back four
years later with his radical proposal to make 10,000 muskets for the U.S. Army
using "machinery moved by water", Wolcott spoke up for him in cabinet
meetings, despite reservations about Whitney's production methods and his
ability to perform on time. Even after his failure to meet the two-year deadline,
the Secretary continued to back him, believing he should have a fair trial to prove
his theories: "I should consider a real improvement in machinery for
manufacturing arms as a great acquisition to the United States." In January, 1801,
his confidence was vindicated. Before Jefferson, Wolcott and other officials,
Whitney demonstrated the practicality of standardization. Picking up parts at
random, he assembled complete gunlocks before their eyes.

David Humphreys was another close friend and business associate. Wolcott
was an incorporator of the Humphreysville Manufacturing Company. Thus, upon
taking up residence again in Litchfield, he was susceptible to his younger
brother's plans for a woolen mill near Torrington. Oliver put up most of the
capital to erect a four-story, red-painted building with the customary churchlike
spire. On the day the roof was raised, the assembled villagers agreed to change
the name of Orleans Village to Wolcottville. In a few years about forty men were
employed in spinning, weaving and dressing fine quality broadcloth. Wolcott
boasted: "About fifty yards of Broad Cloth are manufactured daily, which are
readily (sold) in Boston at a handsome profit, in spite of the competition of
English Traders."

From its beginning, however, the mill had more than its share of money
and management problems. Frederick Wolcott paid more attention to being
judge of probate and a state senator than to seeing that the blue woolen goods
were made profitably. The Wolcotts' manager, James Butler, defrauded the bank,
leaving Oliver responsible for a debt of $40,000 and Frederick with a loss of
$20,000.

Wolcottville

Collinsville, 1876

As elsewhere in Connecticut, the reimportation of cheaper woolens at the end of the war with England had a depressing effect on Wolcottville. The business lost money during most of the years that Wolcott occupied the governor's chair. The new manager, Joshua Clapp, sent the governor a dismal report on operations during 1823–24. On sales of nearly $38,000 the mill lost $8,000. Almost ten thousand yards of cloth were sold, but the price had dropped from as high as eight dollars a yard to less than four dollars. To make Wolcott feel better, Clapp pointed out that all woolen factories were then in the red. From his home the governor wrote that he had no complaint about the manager's conduct or judgment:

> The last eighteen months has been a period of the greatest competition between the British and American Manufacturers of Wool . . . the prices of Cloths are now lower than they were before the revolutionary War and are of much better quality . . .

> The Woolen manufacture is by far the most important of any and the progress of this Country has been such, that it *must and will succeed.* If with our present advantages, we cannot convert Wool into Cloth on as good terms as the Europeans, we must revert to Barbarism . . . Even at present Manufacturers of all kinds are doing as well as Navigators or Farmers and their Capitals are safer than those of Merchants.

Wolcott also felt that the introduction of Lowell's power loom would be of material value to the future of woolen factories. Despite its unprofitability, he felt the Wolcottville enterprise was the equal to any in the country, if not in Europe. In his opinion New England was of one mind on the necessity of protecting manufactures of wool, cotton, iron, leather, wood and paper. "They are now attracting a considerable part of that capital which has heretofore been employed in foreign commerce . . . The actual course of events has united the northern manufacturers with the farmers." He never lost faith in the eventual success of the industry or in a manufacturing economy as the only way for Connecticut to perpetuate its prosperity.

Reverses, however, continued to plague the mill. In 1829 the bank took over the property, selling it back to James Wolcott (a founder) and Samuel Groves for $6,000. During one period of mill idleness the manager inquired of the health of a villager driving by; the latter, pointing to the figure of a ram with great spreading horns atop the mill, replied: " 'Pears to me that the old ram will starve, up there!" He was right. The business barely survived until fire consumed everything in 1844. A new company was formed and a new mill equipped to handle cotton, but its fortunes were no better than its predecessor's. Instead, Wolcottville's future was assured when Israel Coe, in 1834, began to make brass kettles, which grew into Torrington's best-known industry. Wolcottville was then described as a thriving village, containing forty houses, a Congregational church, a brick building used as both a school and for worship by other denominations, four stores, two taverns and a post office.

* * *

Between New Hartford and Unionville the Farmington River at times runs swiftly through one almost continuous gorge until it settles down into a more placid, sinuous flow past the towns of Farmington and Simsbury, on to the Tariffville gorge, and finally down through Bloomfield and Windsor until it empties into the Connecticut. Over the years it provided waterpower for a number of mills and factories including the first American carpet factory (Tariffville, 1825), one of the earliest chair companies (Hitchcocksville, 1818) and the first American axemaker (Collinsville, 1826). In the narrow valley at the sound end of Canton the river broadens, providing an ideal location for the village gristmill and sawmill built much earlier.

After exploring the upper branches of the Farmington River for a manufacturing site with ample waterpower, Samuel W. Collins, a shrewd and energetic young man in his twenties, heard of these mills in "the ravines between the hills" and acquired them. Collins had wanted to go to college. But the death of his father, a Middletown lawyer, when he was thirteen upset his plans and forced him into the employ of his well-to-do uncles, Edward and David Watkinson, Hartford iron merchants. His younger brother, David, soon followed. Both boys married well. Samuel solidified his relationship to Hartford's "Standing Order" by wooing and winning Sarah Colt, daughter of Elisha Colt. David married into the family of William Ely, another affluent merchant.

The Collinses learned all about hardware and determined to start a business of their own. Axes, the indispensable tools of every farmer and frontiersman, particularly interested them. At the time axes were crudely hammered out by local blacksmiths and sold without an edge. This meant the buyer had to spend several hours on grinding. The Collins brothers admired the axes made by one Morgan, a capable blacksmith from Somers who bought his iron from them. David, more daring but less well-balanced, convinced his brother their fortune lay in producing superior-quality ground axes in quantity. Taking into partnership a rich cousin, William Wells, each invested $5,000 and set about converting the South Canton property to their needs. Using a breast waterwheel, trip-hammers and Salisbury iron, eight workmen were soon turning out sixty-four axes a day, forged and tempered, with "light Yankee" heads. Grindstones six feet in diameter and a foot thick were quarried in Nova Scotia, shipped up the Connecticut River to Hartford and then hauled by ox teams fifteen miles west to

Tariffville

Riverton or Hitchcocks-ville

the factory. Always progressive in adopting the latest methods and equipment, the brothers pioneered the use of Lehigh coal in their forges, which replaced the customary charcoal produced from the yellow and white pine woods nearby.

From the start business was so brisk that Samuel Collins had to arrange for transportation and mail service. He persuaded Oliver Couch to divert his four-horse stagecoach from the Albany Turnpike and run it through South Canton on its way to Hartford. The opening of a South Canton post office immediately caused confusion among the postmasters in Canton and North Canton, who urged that the village name be changed to Collinsville, since so many letters were being directed to the "Collins Axe Factory". Oddly enough, Collins himself objected, mainly because he detested the word "ville". He wrote in his diary:

> The name has always been distasteful to me and my family. If I had been consulted and had consented to have my name used it would not have had any *ville* attached to it or been *Frenchified* at all. I would have had it Collinsford, like Torringford and Ashford, which are good Saxon names.

But in 1828 it became Collinsville.

On what he wanted Collinsville to be, Collins had no reservations: "It has been said that our manufacturing villages have a demoralizing tendency. I wish to show there can be an exception. I would rather not make one cent than to have men go away from here worse than they came." Before building his grinding shop, forges and stone office, he put up two boarding houses on Front Street. For many years the upper story of the office served as the only public hall, on Sundays becoming a chapel until the company erected a church in 1836, which Collins himself equipped at a cost of $1,000. The basement made do as a school.

Contrary to prevailing practice, he cut the workday to ten hours and found his men did just as much. Common blacksmiths were hired under a five-year contract starting at fourteen dollars monthly, including board, and rising to twenty-six dollars the last year. A stickler for making the best possible axe, regardless of cost, he discovered that an even greater obstacle to increasing production than the shortage of skilled blacksmiths was the lack of housing in the sparsely settled countryside. In 1831, on the east side of the river, he constructed twenty-one double dwelling houses, all of the same size and plan. Each one rented for $25 a month, while board cost $1.50 a week. To open up more land for houses near the shops, he bridged the canal and river and on the west side added another forty-eight tenements. Evangelical in his religious views, he strongly opposed the Yankee fondness for spirits. To keep Collinsville sober and the local distilleries empty, he bought up two taverns and a drug store. The deeds of the land he sold required the buyer never to use his property for making or selling liquor, upon penalty of forfeiture.

Yet Collins's concern for employee welfare, both moral and physical, was not wholly reciprocated. In a factory notice he complained that too many broke the rules by coming to work late or going home "to haying, and harvesting without leave, and we have not men enough left to carry on our business." In 1833, when his price of twenty dollars a dozen for axes met strong customer resistance and forced him to reduce costs, his workers grumbled about low wages, his withholding of their pay for longer than a year, his temperance principles, and especially the lowering of piece rates for above-average production. Threatening to quit, they formed a committee to bring their grievances to his attention. While they waited outside his office for a reply, he penned a frank, rambling justification of his policies that clearly demonstrated his talent for handling men. Praising their industriousness and dispassionate behavior, he told them:

> I am proud of the fact that we are mentioned far and near as a sample of what manufacturing communities may be in this country . . . Instead of such disorderly and disgraceful conduct as we hear of in manufacturing communities in other countries, we find (freemen) here assembling quietly by hundreds . . . in a truly *Republican town-meeting style.*
>
> I would rather a man would suspect my pecuniary credit than my patriotism or generosity. If there is a favorite object or pursuit with me it is the welfare and happiness of the inhabitants of this village, and that can only be prompted effectually and permanently by such prudent and judicious management of our business as will enable us to meet all our engagements . . .
>
> If you think you can commence work under the new tariff with better courage after a holiday and a game of ball you can take next Monday and enjoy yourselves.

His letter was read before a meeting of the workmen, who then unanimously voted to accept his explanation of the wage cut and to go on cheerfully in the discharge of their duties.

Despite the willing sacrifice of the workers, the Collinses nearly foundered for lack of capital. There was a ready demand for ground and polished axes in the Southern states, but cotton planters, always in debt to Northern merchants, bought on credit. In turn, the brothers had to give the hardware merchants a $33\frac{1}{3}$ percent profit. At the same time they tried to increase production at lower cost by buying machinery to shape and weld the axe poles. The net result was an indebtedness of $250,000. Distrustful of manufacturing, like most banks in those days, the Hartford Bank suddenly demanded immediate payment of its loans. Already heavily mortgaged to William Wells's father, the business had to suspend operations; the property was placed in Wells's hands until a charter of incorporation could be obtained from the General Assembly in 1834 with a capitalization of $300,000. Fortunately, the Bank and other creditors settled their claims by taking stock in the new company—a deal that proved immensely profitable to all in the ensuing years.

Though the spark plug of the Collins Company from its inception, Samuel Collins did not become president until 1845, a position he held until his death in 1871. At the time of reorganization he took the title of superintendent with a salary of $1,500 annually—half what he had earned as a junior partner in his uncle's hardware firm. But he and David also received $50,000 in stock for the exclusive right to use their name on "Edge Tools". Quick to discern talent, Collins attracted men with the same qualities of quiet perseverance and devotion to duty. In his diary, under the year 1832, he made this entry: "E. K. Root commenced work this year as a journeyman machinist in our machine repair shop." Root agreed to work for two years for $546 per annum, to be paid at the end of each year, 312 days to be a year's work. Perhaps the most remarkable mechanical genius to emerge from New England, Elisha K. Root was the young man who befriended Samuel Colt when he worked in his father's silk mill at Ware, Massachusetts, and whom years later Sam would lure away with the offer of a $5,000 salary and the superintendency of his great new armory. While in the service of Collins, Root devised a number of mechanical improvements, some of which he patented, including a machine for punching the eye of the axe out of solid metal and another for sharpening the tool by shaving instead of grinding. His departure, in 1849, must have been a heavy blow to Collins.

For a century the name "Collins" and its trademark, a powerful arm clenching a hammer, were synonymous with "machete", that indispensable agricultural implement and weapon of the frontier. The machete was the outgrowth of bartering in Collinsville axes in which Yankee sea captains engaged as early as 1840. Hartford still functioned as a river port, with vessels sailing to and from the southern trade routes. One day the gleaming steel of Collins axes displayed in a State Street store caught the eye of an enterprising master looking for new goods to trade for sugar and spice. How much better than the crude stone or wooden tools he had seen everywhere in jungle country! They met with a warm reception in South America, but the natives asked also for a reproduction in steel of their own favorite instrument—the oddly shaped machete. The trader brought back a wooden pattern to the Collins brothers, who leapt at the opportunity to create a new product and a new market. That decision in 1845 marked the beginning of a lucrative foreign business, eventually accounting for eighty percent of sales. Eventually, 400 different styles of machete knives with their cow-horn handles were being made from native designs under such export names as "Aquinches", "Coas", "Cavadoras", "Barre-Chancols" and "Arits".

During the 1850s sales averaged close to $400,000 annually. Production rose from 1,200 tools per day to 1,700. A long building was put up to house fourteen more trip-hammers driven by a cast-iron breast wheel twenty-two feet in diameter with eight-foot buckets. While the great carpet concern at Tariffville and Thompsonville temporarily failed along with others, creating greater distrust of manufacturing than ever among Hartford bankers, Collins was earning a fat fifteen percent on its investment, and in 1858 adopted a ten percent annual dividend policy. Samuel Collins could never restrain his contempt for the prevailing credit system and bankers in particular. When the company passed the dividend in 1852 (but never afterward), in order to add $30,000 of profit to surplus for new buildings, the bank became suspicious and insisted on the larger, wealthier stockholders personally guaranteeing its loans. Collins felt that bank directors, when money was tight, took care of only those "who belong to their *clique* and go to the same Church and are political friends. The banks being *short* of funds and more paper being offered they can discount, they all agree to cut down the *large* accounts, especially *Corporations* . . . and the Agent or Manager . . . is told 'you have got rich owners, let them help you.'" He cautioned his successors not to rely on the caprices of bank directors.

Although the workers enjoyed steady employment all year long, wages did not rise proportionately with dividends. They petitioned for an increase in 1846, at a time when the business seemed prosperous, a large stone addition was erected, and more help was being recruited through newspaper advertisements. In a long reply Collins marshaled the arguments one might expect of a besieged manager about competition, falling prices, the need for greater output at lower cost, the risks taken by stockholders, and their normal expectation of at least six percent interest on their investment. Asserting that farming, as the paramount employment, regulated the price of common labor, he told them: ". . . the most feeble man in our employ can earn every month in the year $12.00 to $14.00 easier than he can do it at farming . . .". Farm work, too, required longer hours and greater exertion. "Many of our men quit work in the shops by half past two or by three o'clock and they have plenty of time and strength left for a game of ball, and I am glad to see it, but they ought not to expect us to pay for time spent at play."

Without the steady growth of its foreign business and the stimulus of the California Gold Rush, especially for pick-axes, the Collins Company would have endured rough going. The Southern trade was on the decline. Samuel Collins appreciated the necessity for an aggressive marketing policy. He tried sending out peddlers to demonstrate the superiority of Collins axes. But this proved impractical. Then he sold entirely through hardware merchants in the larger cities, who after a while began to undermine Collins by introducing axes stamped with their own names. Next he experimented with consigning goods to dry goods merchants and paying a five to seven percent commission. They, too, refused to handle Collins axes exclusively. Finally, Collins opened his own sales office in New York in 1848 and hired agents at four percent commission. Imitation of Collins tools, particularly machetes, became widespread. In 1859, the company sued thirty English manufacturers for using its trademark and won; the next year, it succeeded in stopping German infringement.

Over the years the company kept expanding until forty-five separate shops, laid out like a patchwork quilt, spread over eighteen acres on the east side of the river. By 1850 Collins had brought the railroad to the village by donating land for the depot and making a subsidy of $3,000. He always depended on waterpower from the Farmington River, later supplemented by four steam engines. Upstream he also constructed a reservoir that could supply eight million cubic feet of water daily, or two-thirds of his requirements, thus solving the problem of summer droughts. Until late in the century hundreds of tons of Nova Scotian grindstones were still imported each year and worn out. Surrounded by stately elms planted by Collins, the white tenements increased in number to nearly 200, the rents remaining the same. Collins lived long enough to see 600 men make 2,500 tools every working day and his annual sales exceed $850,000.

*   *   *

In retrospect, it seems abundantly clear that the origin of the manufacturing village in New England occurred, not in Rhode Island with Samuel Slater's cotton mill, nor in Massachusetts with Francis Lowell's model community, but in Connecticut with Eli Whitney's gun factory and David Humphreys's woolen enterprise. Slater's pioneering, significant as it was to the development of cotton manufacturing, was primarily confined to the mechanization of spinning yarn. Thanks to his experience in the English mills of Arkwright and Strutt, he appreciated the moral and educational needs of his hands, but only incidentally. Slater did not, at first anyway, conceive of a wholly self-sufficient community, complete with dwellings and churches. In any case, Slatersville came later than Whitneyville. Although he had some philanthropic inclinations, Slater essentially represented the English laissez faire system. His attitudes were not colored by the vision of Puritan utopianism that influenced the actions of the native merchant-capitalists, inventors and reformers.

Connecticut's early founders of manufacturing communities differed greatly from Slater; they resembled more closely Lowell and his Boston friends. Humphreys and Wolcott were gentlemen of culture, wealth and ideals, with a strong urge to reform the domestic economy through both political and business means. Humphreys, in particular, seems to have been the first to have dreamed, with the poet's vision, of a union between farmer and manufacturer that would encourage the one to produce a staple, like wool, that the other could fabricate into a necessity of life, in order to achieve economic self-sufficiency in the new nation. Whitney, the introverted inventor and master machine builder,

Curtisville in Glastonbury

demonstrated a better understanding than Slater of what had to be done to attract and hold skilled mechanics, who were essential to his precision type of manufacturing.

Another difference involved the employment of children. While the apprentice system was universal, there seemed to be a stronger sentiment in both Connecticut and Massachusetts against allowing very young children to work in factories. Certainly, their welfare was better guarded as the result of Humphreys's personal example and his legislative efforts. Aside from the need for apprentices to learn various trades, neither the woolen mills nor the metal-working shops, which eventually dominated the western part of the state, required the large number of child helpers or female hands basic to cotton textile operations.

Entering the scene almost twenty-five years later than Whitney and Humphreys, Samuel Collins was an entrepreneur with democratic proclivities. Beginning as a typical merchant-capitalist, highly religious, a tower of integrity, in many ways he set the stage for the professional executive and the corporate manufacturing organization that developed after 1850. He appreciated the value of having men of ability like E. K. Root; he had a knack for smooth handling of their idiosyncracies; he recognized the necessity of a practical distribution network for his axes and was one of the first manufacturers in New England to build up a lucrative foreign market. Like Whitney, he clearly understood the importance of adequate bank credit for industry on longer terms than the merchant required. Moreover, to a greater degree than any of his predecessors, he devoted himself to the welfare of his employees and to fostering not so much a paternalistic relationship as one based upon mutual respect and understanding, with commonly shared values. He believed that workers benefited as greatly as stockholders, considering the difference in risks. Although no liberal in paying wages, he did introduce the ten-hour workday as early as 1829 and experimented with bonuses for above average output. He demanded that his blacksmiths give as much of themselves as he did, but he was not above talking frankly to them and explaining in detail the problems of his business and the role of its owners. Only in the matter of drinking did he attempt to interfere with their personal lives.

Like Whitney and Humphreys, he inclined to suspicion of any one not a full-blooded Anglo-Saxon American, and employed only those who were orderly, sober and intelligent. Interestingly enough, however, when commencing his first two-story factory, he hired two blacks, the Quicy brothers, who quarried all the stone and laid foundations and walls for several years, in company with "a stout gang of good steady black men." Later, in the mid-forties, when the Irish became the first wave of immigration in Connecticut, he consented to taking on a few of them, but lost patience with their "greenness" and the damage done to his grinding equipment. Nevertheless, Collins lacked any feelings of superiority or class consciousness. Himself unpretentious and disinterested in wealth per se, he advocated a more equal distribution of property "without destroying the inducements to industry and enterprise . . . I suppose that we are all of us in pursuit of *happiness* but I do not believe myself that it is to be found in the accumulation of *wealth*." To his credit, Collins managed his labor relations so well that his employees never struck during his lifetime, and after 1852 the company never had to omit a dividend.

In a sense Collins anticipated the coming of the urban society in which the worker would lose his independence and dignity; instead of a price for his product, which the artisan received, he would have to accept a wage for his labor, thus selling himself rather than his handiwork. The protests of the proud Collins workers in 1833 and 1846 who unconsciously felt themselves becoming wage slaves were not revolutionary but strictly defensive. Instead, capitalism was the radical force, gradually destroying the worker's sense of freedom and equality. Until 1860, the working class movement in the East was merely a sporadic noise, unsuccessful in its attempts to win a ten-hour day in the 1840s, or to establish trade unions in the 1850s. Sumner called this period the "Golden Age" because of its widespread prosperity and the steady progress of industry. It was also one in which the paternalistic manufacturing village gave way to the mill town, although the Cheneys of South Manchester, the Toys and Ensigns of Simsbury, and Samuel Colt of Hartford for many years carried the village to a zenith of development.

CONNECTICUT PLACE NAMES ENDING IN VILLE[1]
Associated with Manufacturing

	Town	Named After	Manufacture	Year Founded
Almyville	Plainfield	William Almy	Cotton goods	1826
Amesville	Salisbury	Horatio Ames	Iron mill	1834
Atwoodville	Mansfield	William & John Atwood	Mansfield Silk Mfg. Co.	1829
Augerville	North Haven	Willis Churchill	Augers	1836
Baileyville	Middlefield	Captain Alfred Bailey	Distillery	1820
Bakersville	New Hartford	Scott Baker	Sawmill and tannery	1804
Ballouville	Killingly	Hosea Ballou	Cotton yarn	1841 (Inc.)
*Bassickville	Bridgeport	Bassick Co.	Automobile hardware	1892
Blissville	Lisbon	Willard Bliss	Cotton warps	1830s
Bozrahville	Bozrah	Town of Bozrah	Cotton & wool fabrics	1814
Bradleyville	Litchfield	C. S. Bradley	Iron mill	1810
Bradleyville	Middlebury	L. Bradley	Knives	1866
*Bradleyville	Weston	G. W. Bradley & Sons	Edge tools	1892
Burrville	Torrington	John M. Burr	Grist & shingle mill	1829
Burtville	Derby	David Burt	Axe helves & hoe handles	1857
Centerville	Vernon		Cotton goods	1809
Chafeeville	Mansfield	O. S. Chaffee	Silk	1838
Chapinville	Salisbury	Chapin family	Iron furnace	1800
*Cheneyville	Manchester	Cheney family	Silk	1838
Clarksville	Stonington	Clark Thread Mill	Cotton thread	1891 (Inc.)
Clayville	Griswold	Senator Henry Clay	Cotton goods	1828
Clintonville	North Haven	David L. Clinton		1853
Collinsville	Canton	Collins brothers	Axes	1826

Village	Town	Proprietor	Product	Date
Conantville	Mansfield	Captain Josiah Conant	Silk thread	1829
*Converseville	Stafford	Eliot & Parley Converse	Satinet	1840
Coreyville	Lebanon	Joseph Corey	Cotton goods	?
*Curtisville	Glastonbury	Frederick Curtis	Silver plating	1854
Daleville	Willington	Thomas Dale	Silk	1840
Danielsonville	Killingly	James Danielson	Cotton goods	1810
Daytonville	Torrington	Jonah Dayton	Sawmill & pipe organs (1840)	1809
Dayville	Killingly	Captain John Day	Cotton goods	1832
Doaneville	Griswold	Joseph H. Doane	Cotton cloth	1823
Dobsonville	Vernon	Peter Dobson	Cotton spinning	1810
Donkeyville	Tolland	Donkeys used to haul cotton	Cotton mill	1800s
*Dorrville	Griswold	Thomas W. Dorr	Twine	?
Eagleville	Mansfield	Eagle Mfg. Co.	Cotton & wool goods	1822
*Eliottville	Killingly	Elliottville Mfg. Co.	Cotton fabrics	1856
Elmville	Killingly	Elmville Mfg. Co.	Sash cords & mill supplies	?
*Elmville	Norwich	Whitestone mills	Dyeing of woolen goods	?
*Fisherville	Thompson	Fisher Cotton Mill	Cotton	?
Fitchville	Bozrah	Asa Fitch	Iron works	1828
Floydville	East Granby	Marcus W. Floyd	Shade tobacco	?
Fluteville	Litchfield	Asa Hopkins	Applewood flutes	1830
Forestville	Bristol		Wooden clocks	1833
Foxville	Stafford	C. Fox	Cotton fabrics	1837
*Fraryville	Meriden	Frary & Benham	Britannia ware	1840
Gaylordsville	New Milford	Gaylord family	Gristmill	1807
Glenville	Greenwich	(descriptive)	Woolens	1854
Glynville	Stafford	Glyn Mills	Cotton warps	1857
Grantville	Norfolk	Grant family	Sawmill	?
Grayville	Hebron	William Gray (Hebron Mfg.)	Cotton twine	1814

	Town	Named After	Manufacture	Year Founded
Greeneville	Norwich	William P. Greene	Cotton goods	1826
Greenmanville	Stonington	Greenman family	Shipyard	1827
Griswoldville	Wethersfield	Griswold family	Carding and fulling mill	1680
			Textiles	1832
Gurleyville	Mansfield	Ephraim Gurley	Screw auger & blacksmith	1814
			Silk mill	?
Hallsville	Preston	Hall brothers	Woolens	1848
*Hallville	Willington	Hall family	Cotton sewing thread	1828
Harrisville	Woodstock	Captain Edward B. Harris	Cotton fabrics and twine	?
Haywardville	East Haddam	Nathaniel Hayward, India rubber manufacturer	(none)	
Hazardville	Enfield	Col. Augustus G. Hazard	Powder	1835
Hilliardville	Manchester	Elisha E. Hilliard	Woolen goods	1825
*Hitchcocksville	Barkhamsted	Lambert Hitchcock	Chairs	1818
Hoadleyville	Plymouth	Silas Hoadley	Clocks	1810
Hopeville	Griswold	Hopeville Pond	Woolen mill	1818
Hopeville	Waterbury	Hope Mfg. Co.	Metal wares	1851
Hotchkissville	Woodbury	Hotchkiss family	Shears and Knives	1851 (Inc.)
*Hoytville	Stamford	Hoyt, Lyman Sons	Mattresses	1892
*Humphreysville	Seymour	Col. David Humphreys	Woolen mill	1806
Huntsville	Canaan	Hunt family	Iron furnace	?
Hydeville	Stafford	Nathaniel Hyde	Blast furnace	1796
Johnsonville	East Haddam	Emory Johnson	Twine	1832
Joyceville	Salisbury	Joyce family	Charcoal furnace	1765
*Kelloggville	Vernon	Nathaniel O. Kellogg	Textile mill	?
Kenyonville	Woodstock	A. & W. Kenyon	Woolen fabrics	?

Place	Town	Named for	Product	Date
Lakeville	Salisbury	(descriptive)	Blast furnace	1775 (?)
Lanesville	New Milford	Jared Lane	Gristmill & iron works	1716–1755
Laysville	Old Lyme	Lay Factory	Woolen fabrics	?
Leesville	East Haddam	Henry S. & S. H. Lee	Cotton sheetings	1826
Lydallville	Manchester	Henry Lydall	Knitting-machine needles	?
*Masonville	Thompson	Mason family	Cotton goods	1813 (Inc.)
*Mechanicsville	Thompson	(descriptive)	Cotton goods	?
Merrowville	Mansfield	Joseph M. Merrow	Knitted stockings	1838
Merwinsville	New Milford	Sylvanus Merwin	Store and hotel	1827
Mixville	Cheshire	John Mix		1853
Montville	Montville	(first ville in Conn.)		1786
Mooreville	Winchester	Franklin Moore	Bolts and nuts	1867 (Inc.)
Newhallville	New Haven	George T. Newhall	Carriages	?
Northville	New Milford	(descriptive)	Paper mill	1852
*Oakville	Watertown	(descriptive)	Pins and wire goods	1852 (Inc.)
Orcuttville	Stafford	Orcutt brothers	Saw, oil & carding mills	1800s
Packerville	Canterbury	Daniel Packer	Cotton	1818
*Packwoodville	Colchester	Packwood family	Gristmill	?
Phoenixville	Eastford	Phoenix Mfg. Co.	Cotton batting & twine	1831
Pineville	Killingly	Pineville Mfg. Co.	Shoddy	?
Plainville	Plainville			1869
Plantsville	Branford	Raymond Plant	Seeds	?
Plantsville	Southington	A. P. & E. H. Plant	Bolts and nuts	1853 (Inc.)
*Prattsville	Meriden	Julius Pratt	Ivory combs	1822
Rhodesville	Putnam	James Rhodes	Cotton fabrics	1820
*Richmondville	Westport	David Richmond	Cotton & woolen goods	1814
Robertsville	Colebrook	Clark Roberts (postmaster)	Iron furnace	1775
Rockville	Vernon	(descriptive)	Textile mills	1794

	Town	Named After	Manufacture	Year Founded
Rossiterville	Torrington	Newton Rossiter	Tannery	1818
Sabinville	Sterling	Henry Sabin	Cotton mill	1830
Shailerville	Haddam	Shailer family	Brownstone quarry	1788
*Smithville	Derby	Sheldon Smith	Wire and nails	1833
*Smithville	Windham	Smithville Mfg. Co.	Cotton thread	1857 (Inc.)
Somersville	Somers	John Somers	Satinet	1836
Spoonville	East Granby	(descriptive)	Silverplating of spoons	1840
Springville	Vernon	(descriptive)	Cotton & cassimeres	1821
Staffordville	Stafford	Town of Stafford	Iron foundry	1830
Stillmanville	Stonington	O. M. Stillman	Woolen mill	1831
Storrsville	Mansfield	Storrs family	Cotton & woolen goods	1818
Taftville	Norwich	Cyrus & Edward P. Taft	Cotton, rayon & velvet	1865 (Inc.)
Talcottville	Vernon	Horace W. & Charles Talcott	Cassimeres	1854
Tariffville	Simsbury	Tariff law of U.S.	Carpets	1825
Terryville	Plymouth	Eli Terry	Clocks	1835
Thamesville	Norwich	(descriptive)	Woolen goods	?
Thompsonville	Enfield	Orin Thompson	Carpets	1828
*Toyville	Simsbury	Joseph Toy	Safety fuse	1836
Tuckerville	Stafford	Mr. Tucker	Satinets	1858
*Turnerville	Hebron	Phineas W. Turner	Silk	1853
Uncasville	Montville	Indian sachem Uncas	Cotton & woolen fabrics	1799
Unionville	Farmington	Union Nut Co.	Bolts & nuts	1864 (Inc.)
Unionville	Norwalk	Union Mfg. Co.	Felt cloth	1838 (Inc.)
Unionville	Plainfield	Union Cotton Mfg. Co.	Cotton	1811
*Walterville	Bridgeport	Edward P. Walter	Machinery	?

Waterville	Waterbury	(descriptive)	Knitted woolen goods	1816 (Inc.)
Wellesville	New Milford	George Welles	Satinets	?
Wellesville	Windham	?	?	?
Westville	Danbury	(descriptive)	Hats	?
Westville	New Haven	(descriptive)	Various	?
Whigville	Burlington	Whigs	Copper mine	1839
Whitneyville	Hamden	Eli Whitney	Guns	1798
Williamsville	Killingly	Caleb Williams	Cotton mill	1827
Wilsonville	Thompson	O. S. Wilson	Woolen goods	?
Windsorville	East Windsor	(descriptive)	Cotton fabrics	1829 (Inc.)
Wolcottville	Torrington	Oliver Wolcott Jr.	Woolen mill	1813
Wrightville	Torrington	Robert Wright	Scythes	1852 (Inc.)
Yalesville	Wallingford	Charles & Hiram Yale	Tinware & Britannia ware	1809
*Youngsville	Barkhamsted	Joshua Youngs	Sawmill	1800

[1] Compiled in cooperation with Dean Arthur H. Hughes of Trinity College.

* No longer in existence.

[CHAPTER III]

The Factory System

FREE enterprise and the capitalist system, it is clear, were not New England
inventions, since they evolved out of English mercantilism. Yet, with unique
resourcefulness the Yankees did expand and refine them to the dominant
economic concept in nineteenth-century America. The prodigious offspring of
capitalism, the factory system, was an entirely different matter. Admittedly, our
manufacturing development owes much to the technical progress achieved in
Europe in the late 1700s in power spinning and weaving, the steam engine, coal
and coke for smelting iron, chemicals for dyeing and the rolling mill. The
contributions of men like Wyatt, Arkwright, Crompton, Cartwright, Watt,
Maudslay and Wilkinson to the Industrial Revolution cannot be underestimated.
Nevertheless, it took the native genius of the Yankee to apply, adapt and
combine English mechanical inventions so as to create the radically new and
infinitely versatile idea of interchangeable parts and mass production. This
stupendous development over a mere twenty-year period, beginning with Samuel
Slater and culminating with Eli Whitney, made the American factory system
dominant over the traditional household industries by 1830.

Just what distinguished the factory from the colonial mill or shop? The
essential difference lay in machinery and management. As Victor Clark points
out: "It is not a question of specialization, nor ownership, nor completeness of
process, nor size, but rather depends upon a combination of equipment and
organization." To which might be added the human element of a few
independent artisans or craftsmen versus a group of paternalized employees.

As in England, the initial steps toward factory organization in America
centered around the spinning of cotton yarn, as Slater did, by assembling a
number of workers under one roof. Weaving yarn into cloth, however, remained
for several decades a household function. Besides a supply of cheap labor, an
abundance of water was vital to the factory system. Alexander Hamilton deserves
credit for the first concerted attempt to apply waterpower extensively for

manufacturing, with the incorporation of the Paterson mill on the falls of the Passaic River in 1791. Another milestone was the introduction at Waltham in 1814 of Lowell's power loom. By contrast, England had mechanized weaving eight years earlier. But, Lowell's concentration of all the operations in a single plant made his the first integrated cotton mill in the world and represented a significant advance toward modern factory production. The Boston Manufacturing Company also combined the other essential ingredients for successful factory organization: ample capitalization ($600,000), standardized production (Waltham specialized in sheetings and shirtings), division of labor into departments, a single marketing agency instead of jobbers and commission agents and supervision by men chosen for their executive ability rather than mere technical know-how.

For fifteen years following Slater's achievement, cotton processing in New England made little headway. New enterprises were numerous, but most did not survive long. Of eighteen mills started between 1790 and 1805, six were in Connecticut. Then, a combination of fortuitous circumstances suddenly provided the impetus for rapid and widespread growth not only in textiles, but in powder, paper, iron goods, guns, clocks, tinware, hats, buttons, plows, wagons and carriages, leather goods and liquor. Farm people moved to the new manufacturing villages. A class of wage earners emerged. Nevertheless, the population remained primarily rural in character until 1850, with factories scattered around the countryside rather than clustered in the larger towns, as they were after the Civil War.

Contributing to this development were, of course, the availability of cheap power and surplus capital, the adaptability of most Yankees and the patriotic desire to be independent of foreign goods. But the foremost factor was the Embargo of 1807–09, which halted trade with England. Through the Embargo President Jefferson hoped, by prohibiting both exports and imports, to end England's insulting practice of impressing American sailors into the British Navy. It proved to be the outgoing President's worst mistake. It ruined American shipping, idled sailors and loaded the jails with debtors. In New Haven harbor seventy-eight vessels were tied up; hundreds of sailors, shipwrights and merchants lounged about the wharves, cursing the "Dambargo". The Federalists, now in their political twilight and still worshiping the maritime trade, bitterly opposed the Embargo, much to the disgust of pro-manufacturing Republicans. It affected England not at all and failed to protect American ships or seamen.

It was followed, in 1809, by the Non-Intercourse Act, which to some extent reopened trade, and by a promise to stop trading with the enemy of the first power, England or France, that recognized our neutrality. Connecticut Republicans deplored the inaugural address of the Federalist governor John Treadwell in

Old Slater mill, Pawtucket, 1792

Textile mill at Norwich Falls

May, 1810, because they read "not a word of manufactures, although they are more formidable to Britain than a navy of 100 ships of the line."

England's procrastination, despite a willingness to settle the dispute short of conflict, led to the War of 1812, pushed by such young war hawks as Henry Clay and John Calhoun, who had had their fill of diplomacy and economic sanctions. In the face of the united opposition of New England congressmen, they aroused the country to fight for "Free Trade and Sailors' Rights", although their real motives seemed directed, not toward the sea, but more toward the enticing prospects of conquering Canada, ending the Indian menace and opening up more western land for settlement. The War dealt a death blow to maritime interests, as well as to many farmers and merchants, yet since it eventually united the country and gave the equivalent of maximum protection to domestic industry, it served as the catalyst for accelerating Connecticut's rise as a manufacturing state.

To a large extent Connecticut's cotton manufacturing was the outgrowth of Rhode Island pioneering. Pawtucket spinners, like the Slaters and Wilkinsons, moved into such nearby towns as Killingly, Plainfield, Putnam, Thompson and Windham. Within these townships, on the streams flowing into the Thames River at Norwich, were clustered more than seventy percent of all cotton spindles in the state. Catching the fever, local farmers invested their meagre savings in stock companies for weaving cotton cloth. The most ambitious undertaking was the Pomfret Manufacturing Company, organized in 1806 by the Wilkinsons to be a model factory, with salubrious working and living conditions. So rapid was the spread of the cotton industry throughout eastern Connecticut, with Windham County the center, that in 1811 the Windham *Herald* was moved to ask: "Are not the people running cotton mill mad?"

Meanwhile in Vernon, Peter Dobson made local history by erecting the town's first cotton mill and introducing a new kind of cloth. He had worked a short while with Samuel Slater and, like him, was an expatriate from England, who had daringly circumvented his homeland's anti-emigration laws by concealing himself in a hogshead and being rolled aboard ship. The war with England gave him an opportunity to make material for uniforms. By chance shown a piece of satinet, half cotton and half wool, he soon produced this style in quantity.

Manufacturers who failed or ended up heavily in debt faced the prospect of prison. Two such unfortunates were George and Richard Lord of Leesville, who ran a woolen business and sawmill. In 1815, fire consumed their buildings, causing a loss of $25,000. Undaunted, they rebuilt the next summer, this time using brick, and added another 500 spindles. Suffering heavy losses, they had to mortgage their property to the state because of loans received from the School Fund. Foreclosure followed in 1822; both were arrested and jailed, but Richard contrived to escape and run away to Ohio.

In 1810, the woolen industry in Connecticut was also well established, although it developed much more slowly owing to the lack of machinery for weaving. Of fifteen such mills, five were located in New London County. The Republican party called for support of the industry on the grounds that Americans should be freed from wearing "foreign livery." Under the Embargo prices rose to nine and ten dollars a yard, and the demand for the Merino wool introduced by Colonel Humphreys remained high despite the great increase in sheepherds. The spread of woolen mills apparently had no adverse effect on household spinning—not for a decade or two, anyway. This was the year the Humphreysville Company received its charter with the right to sell $500,000 in stock at $400 per share. Although still weaving cloth by hand, its equipment shared some features of the factory system. It was the country's best woolen mill. As guest of honor at a dinner for manufacturers in Philadelphia, in 1808, Humphreys had toasted domestic textiles with this rhetorical boast: "The Best mode of Warfare for our Country—the artillery of carding and spinning machinery, and the musketry of shuttles and sledges."

The Scholfield family made a contribution to the woolen industry second in importance only to that of Humphreys. In fact, these gifted artisans have been called the fathers of the American woolen mill, as Slater is considered the originator of the cotton factory. Just as Slater evaded British laws and brought to America the knowledge of cotton machinery, so Arthur and John Scholfield left Yorkshire in 1793 to build a string of factories in Massachusetts and Connecticut. Their mill at Byfield holds the honor of being the first to use all power-driven machinery, but the Hartford Woolen Manufactory properly claims precedence in wool spinning by hand. On one of his wool purchasing trips, John became interested in a waterpower site at Montville, Connecticut, and there the brothers settled in 1799. This operation represented the first complete power-operated woolen factory in the state. A third brother, James, carded wool at North Andover, Massachusetts. In 1801, Arthur broke up the partnership because of a dislike for his sister-in-law and moved to Pittsfield, where he put together carding machines. John stayed in Montville until 1806, then sold out and bought a mill site at Stonington. Seven years later he purchased another mill in Montville and the next year set up a plant at Waterford under the management of his son Thomas. Another son, John Jr., operated a wool carding shop in Jewett City, charging twelve cents per pound for this service, and when he added fulling it became a full-fledged woolen mill by 1816. Incidentally, Samuel and John Slater came to Jewett City in 1826 and started a cotton mill with ninety looms. John Scholfield's mills were all small but so thriftily managed that upon his death in 1820 they were debt-free. The Scholfields' greatest achievement was their dissemination throughout New England of the know-how for an improved wool

Brandegee's thread mill, Berlin

carding machine, an operation fundamental to the making of good quality cloth and also to the perfection of the factory system for woolens.

The earliest known use of steam for manufacturing occurred in 1811 with the installation of an Oliver Evans engine at the Middletown Manufacturing Company, the fourth woolen mill established in Connecticut, following the Hartford Woolen Manufactory, Montville and Humphreysville. This too was a substantial enterprise, chartered with a capitalization of $200,000. The factory was located in a five-story brick building, formerly a sugar house, with a separate dye house. As many as eighty hands were employed at the peak of operations: they produced about $70,000 worth of woolens annually, which sold for nine to ten dollars a yard.

The principal founders were Arthur Magill, a prosperous Middletown merchant, his son and Isaac Sanford, an English mechanic with steam engine experience. As an importer Magill suffered from the Embargo and the consequent scarcity of woolen goods, and like so many other Connecticut merchants he decided to put his capital into manufacturing. The Evans engine, 24-horsepower, cost the partners around $15,000. Sanford was enthusiastic about its performance, considering it superior to the Watt's engines on which he had worked. It drove all the machinery for carding, spinning, reeling, weaving, washing, fulling, dyeing, shearing, dressing, and finishing, as well as heating the building. The

Middletown Company continued in business for a decade, finally succumbing as the result of legal entanglements.

Another Middletown enterprise carved out a special niche for itself in the textile industry that endured until modern times. This was the Russell Manufacturing Company, founded in 1834 by Samuel Russell, Samuel D. Hubbard, George Spaulding and other businessmen of that town. Samuel Russell left his birthplace as a young man to make a reputation and a fortune in foreign trade. In Canton, China, he established Russell & Company, one of the four American houses to survive the intense competition in the Far East. His partners included several well-known Massachusetts merchants, particularly Augustine Heard, the distinguished Ipswich sea captain who later started his own trading house. In the Canton "factory", as the trading centers were called, isolated from the Chinese community, Russell lived a hermitlike life for a quarter of a century. Returning to Middletown, he erected an elegant mansion and assumed a leadership role. Under his direction, Russell Manufacturing, in 1841, began to weave elastic webbing on hand looms. With the help of a Scotch expert, the company soon shifted to machine-operated looms that insured preeminence in its field. As president of the local bank, Russell rescued it during the panic of 1857 by advancing $75,000 of his own resources.

Tench Coxe's survey of manufactures revealed the leading pursuits in Connecticut as of 1810.

Industry	Dollar Value
Woolen Goods (made in families)	$1,098,242
Distilleries (560)	811,194
Linen Goods (made in families)	800,359
Hats	522,209
Tanneries (408)	476,339
Blended Cotton Coths, Yarn & Stockings	352,243
Rope Walks (18)	243,950
Saddlery & Shoes	231,812 ·
Furnaces & Forges (56)	230,090
Tin Plate	139,370
Wooden Clocks	122,955
Buttons	102,125
Scythes & Axes	91,145
Paper (19)	82,188
Combs	70,000
Carriages (4)	68,855
Flaxseed Oil Mills	64,712
Glass & Pottery (14)	58,100

Brass, Jewellery & Plated Ware   49,200
Guns (3)   49,050
Powder Mills (7)   43,640

The grand total: almost $6,000,000. Obviously, although Coxe specifically listed fourteen cotton and fifteen woolen manufacturing establishments, each employing between fifteen and twenty workers, most of the spinning and weaving of cotton, linen and woolen goods was left to the housewife. At this time the United States had only 87 cotton establishments; by 1831 the number had jumped to almost 800, two-thirds of them in New England.

The majority of the tanneries and distilleries were one-man enterprises. Distilling, as the second largest industry, was centered around Enfield and the Windsors. The *Gazetteer* of Pease & Niles called it the most important and extensive manufacturing interest in Hartford County. Almost 200 distilleries

STATE OF CONNECTICUT—MANUFACTURES.

COUNTIES.	Cotton Manufacturing Establishments.	Flaxen Goods in Families, &c.		Hempn Manufacturing Establishments.	Blended and unnamed Cloths and Stuffs.		Woollen Goods in families, &c.	
		Yards.	Value in Dollars.	Value in Dollars of Goods Made.	Yards.	Value in Dollars.	Yards.	Value in Dollars.
Hartford,	5	390,169	133,301 99	.	7,234	2,745 24	188,663	193,311 45
New-Haven,	2	292,561	89,886 45	.	64,864	26,772 90	131,054	141,676 75
New-London,		266,248	90,524 32	9,148 40	167,188	70,829 44	114,760	83,683 4
Fairfield,	1	412,006	136,867 15	1,750	10,054	4,858	159,572	157,229 74
Windham,	4	253,582	87,689 52	.	291,980	112,756 71	109,852	86,688 50
Litchfield,		431,194	157,129 24	.	16,700	5,678	281,184	278,496 68
Middlesex,	1	150,839	48,697 28	1,250	42,655	15,582 70	67,662	85,406 76
Tolland,	1	165,479	56,262 86	.	5,000	2,000	86,998	71,749
Total amount,	14	2,362,078	600,358 81	12,148 40	605,675	241,522 99	1,119,145	1,098,241 92

COUNTIES.	Woollen Manufacturing Establishments.	Stockings and Web Suspenders.	Sewing Silk and Raw Silk.	Looms for Cloths of Cotten, Wool, &c.		Carding Machines.		Fulling Mills.	Spindles.	Hats.		Blast and Air Furnaces.	
		Value in Dollars.	Value in Dollars.	Number.	Number.	Pounds.	Number.		Number.	Value in Dollars.	Number.	Value in Dollars.	
Hartford,	2	33,302	.	2,372	35	73,419	39		2,014	5,000			
New-Haven,	1	1,600	.	1,566	28	76,500	33		2,558	45,400			
New-London,	5	8,874	384	2,240	19	79,999	19		514	29,350	1		
Fairfield,	2	21,483½	.	1,897	36	101,200	35		252	348,791			
Windham,	1	.	27,375	2,435	17	64,470	21		5,477	14,490			
Litchfield,	3	37,762	.	3,279	30	85,000	45		250	45,707	4	30,500	
Middlesex,	1	.	.	1,101	10	20,000	14		400	22,961	1	2,240	
Tolland,		8,000	744	1,242	9	3,500	12		418	7,530	2	13,440	
Total amount,	15	111,021½	28,503	16,132	184	504,088	218		11,883	522,209	8	46,180	

STATE OF CONNECTICUT—MANUFACTURES.

COUNTIES.	Forges.			Trip Hammers.		Rolling and Slitting Mills.	Nails.		Gun Smiths.		Tin Plate Work.	Type Founderies.	Brass, Jewellery and plated ware.
	Forges.	Tons of Bar Iron.	Value in Dollars.	Number.	Value in Dollars of Work.		Naileries.	Value in Dollars of Nails.	Guns.	Value in Dollars.	Value in Dollars.		Value in Dollars.
Hartford,	3	134	20,580	2	1,680	.	3	3,510	1,400	12,300	57,690	1	18,000
New-Haven,	4	7	980	.	3,150	.	.	.	2,000	26,000	57,080	.	6,400
New-London,	3	58	7,460	5	$15,825\frac{42}{110}$	.	3	3,240	.	.	.	.	6,900
Fairfield,	2	45	10,175	3	8,500	1	5	6,260	.	.	.	.	4,000
Windham,	1	5	800	12	13,400	.	1	272	.	.	.	.	.
Litchfield,	32	1,164	139,475	8	38,400	2	4	6,410	.	.	19,000	.	7,000
Middlesex,	1	15	1,800	2	1,340	.	1	5,600	1,000	10,750	5,000	.	4,700
Tolland,	2	22	2,640	.	9,550	.	1	1,800	.	.	.	.	2,200
Total amounts,	48	1,450	183,910	32	$91,145\frac{42}{110}$	3	18	27,092	4,400	49,050	139,670		149,200

COUNTIES.	Brass Founderies.	Buttons.		Tanneries.		Saddlery, Shoes and Shoe-Binding of Leather.	Flax Seed Oil.		Spirits Distilled.		
		Groce.	Value in Dollars.	Number.	Value of Leather in Dollars.	Value in Dollars.	Mills.	Value in Dollars.	Distilleries.	Gallons distilled from Fruit and Grain.	Value in Dollars.
Hartford,	2	26,000	13,000	65	88,246	.	10	19,123	198	727,765	447,362
New-Haven,	1	129,000	89,125	93	100,972 50	12,800	.	.	54	104,735	57,897
New-London,	.	.	.	40	54,124 95	.	3	20,264	10	36,650	20,325
Fairfield,	.	.	.	45	71,311 66	65,112 50	.	.	116	62,594	36,461
Windham,	.	.	.	32	34,894 50	6,500	5	5,000	22	38,270	19,135
Litchfield,	.	.	.	62	75,869 75	90,477 50	2	10,500	103	199,890	102,363
Middlesex,	1	.	.	52	38,769 50	38,322	2	6,625	10	169,400	109,450
Tolland,	.	.	.	19	12,150	18,600	2	2,500	47	35,100	18,150
Total amounts,	4	155,000	102,125	408	476,338 86	231,812	24	64,712	560	1,374,404	811,144

H

produced 728,000 gallons annually, but the *Gazetteer* singled out only twenty-one as having any real size, indicating that most were family affairs capable of making only a few gallons apiece on a daily basis. Whereas in colonial times these crude stills, housed in shacks with dirt floors, turned sugar and molasses into rum, now they concentrated on cider brandy and gin. Only a few years earlier large-scale continuous stills had been devised by the British and French, and the Yankees took advantage of the indigenous Indian corn or maize to prepare the mash of three parts corn and one part malt and rye. Then the mash was fed into the still, the alcohol boiled, and the refined neutral spirits siphoned off, free from odor or taste of grain. The waste was fed to pigs and cattle in order to fatten them for export. Distilling was generally done in the fall, when temperatures most favored fermentation.

STATE OF CONNECTICUT—MANUFACTURES.

COUNTIES.	Carriages. *Value in Dollars.*	Wooden Clocks. *Number.*	*Value in Dollars.*	Paper. *Mills.*	*Value in Dollars.*	Marble Work. *Value in Dollars.*	Glass. *Works.*	*Value in Dollars.*	Potteries. *Number.*	*Value in Dollars of Wares.*	Rope Walks. *Walks.*	*Value in Dollars of Cordage.*
Hartford,	.	5,565	38,955	6	17,820	.	2	27,360	9	9,740	3	33,150
New-Haven,	14,080	5,000	50,000	4	24,780	7,500	.	.	.	.	2	36,000
New-London,	.	•	.	3	13,980	.	.		.	.	6	122,300
Fairfield,	14,275	.	.	2	14,100	.	.		3	21,000	2	
Windham,	.	.	.	1		.	.		.			
Litchfield,	23,000	4000	34,000	1	5,508	3,500						
Middlesex,	.	.	.	1	2,000	.	.		.	.	5	52,500
Tolland,	17,500	.	.	1	4,000		.					
Total amounts,	68,855	14,565	122,955	19	82,188	11,000	2	27,360	12	30,740	18	243,950

COUNTIES.	Gun Powder. *Mills.*	*Value in Dollars.*	Combs. *Value in Dollars.*	Straw Bonnets. *Value in Dollars.*	Miscellaneous Goods. *Value in Dollars.*
Hartford,	7	43,440	15,000		
New-Haven,	.	.	7,000	2,000	39,900
New-London,	.	•	.	.	22,352
Fairfield,	.	.	7,000	.	9,360
Windham,	.	200	6,500	17,000	
Litchfield,	.	.	1,500	1,800	
Middlesex,	.	.	33,000		
Tolland,	.	.	.	6,300	
Total amounts,	7	43,640	70,000	27,100	71,612

With six gin distilleries, Warehouse Point alone bottled more spirits from grain than any other place along the Eastern seaboard, paying almost $24,000 a year in duties. Despite Puritan morality and the never-legislated Blue Laws, Yankees possessed a thirst as unquenchable as their energy and ingenuity. Pease & Niles acknowledged the moral concern but found good reason for justifying the industry's existence on economic grounds. The Connecticut Temperance Society, organized in 1829, disagreed. Its members reported that the state had over 1,400 retailers of spirits, one for every twenty-five families, and estimated that consumption per person averaged up to five gallons a year. Based on actual production figures, this estimate was quite accurate. Dr. Lyman Beecher expressed his disgust at the tippling, however moderate, that he observed at gatherings of Connecticut ministers. Clergymen, as leaders of the "Standing Order", merely reflected—and countenanced—the Anglo-Saxon tradition of drinking to ameliorate the everyday life of hardship, exposure and disease which most colonists had to endure. Besides, there was a prevalent folk belief in the medicinal properties of alcohol. By 1845, the seventy distilleries remaining in Hartford County were making only one-third as much liquor, almost 52,000 gallons of gin and an equal amount of cider brandy annually. But it was less the reform efforts of the temperance advocates than the inroads of foreign trade that, starting in 1816, depressed this profitable business. Eventually the pungent stills ceased to occupy the labors of Yankee entrepreneurs, while the imbibing habits of the majority continued unabated.

* * *

From boyhood Eli Whitney had shown an exceptional mechanical aptitude for repairing such intricate things as violins and watches and for using all kinds of tools on his father's Massachusetts farm. For spending money he made nails, hat pins and walking canes. Finally realizing he would never amount to more than a clever mechanic without a college degree, he taught school long enough to pay his own way and at twenty-three years of age was admitted to Yale. After graduation he conceived, while studying law in Savannah, Georgia, a machine to clean the seed from green, short staple cotton, a process that took a slave ten hours for just one pound of lint. This was the cotton gin, which according to Tench Coxe saved manual labor in the proportion of 1000 to one. It made cotton king in the South until the Civil War and stimulated the expansion of cotton manufacturing in the North.

Six years after his invention of the cotton gin, Whitney offered to make the government thousands of muskets from interchangeable parts, a method never before tried in the United States (at least on guns), although he had no prior gunmaking experience. Gunmaking in 1798 was strictly a handicraft, slow and unsystematic. Without a clear vision of fathering what he called "an entirely new and different system" of manufacture, Whitney certainly would never have dared to submit such a grandiose proposal. Disillusioned by interminable legal wrangling over the infringement of his cotton gin patent, the inventor decided then and there never to seek protection again for the original methods, tools and machines he was about to create. The results became his crowning glory, achieved singlehandedly, perhaps with some inspiration from similar but less spectacular efforts in Europe, but without knowledge of the remarkably parallel and nearly simultaneous cerebrations of another gunmaker a few miles distant.

The influence of Whitney's friend, Oliver Wolcott Jr., succeeded in his getting a contract to produce 10,000 French-type muskets for a price of $13.40 each. Twenty-seven other gunmakers also received contracts at the same time. Fourteen months later, having found a millsite with an excellent waterfall just

Eli Whitney

Plaque at Whitneyville site

across the New Haven town line in Hamden and overcoming numerous obstacles to build, equip and man an armory, Whitney wrote Wolcott:

> One of my primary objects is to form the tools so the tools themselves shall fashion the work and give to every part its just proportion—which when once accomplished will give expedition, uniformity and exactness to the whole . . . In short, the tools which I contemplate are similar to an engraving on copper plate from which may be taken a great number of impressions exactly alike.

By uniformity he meant a tolerance of about $\frac{1}{32}$ of an inch, far inferior to the rigid machine tolerances of modern metalworking. Because of the dearth of experienced workmen he stressed the fact that should the execution of his plan fail, each mechanic would have to form every part "according to his own fancy & regulate the size & proportion by his own Eye," thus resulting in nonconforming parts and requiring triple the number of hands.

His first innovation was probably a filing jig to hold the pieces of the musket accurately. After two years of delays and disappointments, at the end of which time he could deliver only 500 guns, he apparently had in operation, besides his forge, trip-hammer and new-fangled jigs, machines for drilling parts and boring barrels. Sometime between 1801 and 1807 his system came to fruition. He completed his first contract almost entirely with unskilled labor, while several other gun contractors, depending wholly on skilled craftsmen, failed. Eventually, on one of his trips to Washington he gave a convincing demonstration to the Secretary of War of his standard, machine-made parts by assembling ten muskets from a pile of parts scattered on the floor. The government arms inspector, Captain Decius Wadsworth, a frequent visitor and friend, claimed that Whitneyville outperformed the Springfield Armory with much less manual labor, and he had nothing but the highest praise for Whitney's ingenuity and integrity: "Patient, prudent, of mature reflection, diligent, economical, blest with sound judgment, it is rare to find a man uniting so many excellencies, free from striking defects."

President Timothy Dwight of Yale described the operations at Whitneyville as singular:

> In forming the various parts of the musket, machinery, put in motion by water, and remarkably well adapted to the end, is used for hammering, cutting, turning, perforating, grinding, polishing, &c. The proportions and relative positions of the locks are so nearly similar, that they may be transferred from one lock and adjusted to another . . . By an application of the same principles, a much greater uniformity has also been given to every part of the muskets . . .

By 1815, Whitney's system had won widespread recognition in America.

Basic to the development of mass production has been the use of powered

Whitney's Charleville musket, .69 calibre, made after 1801

metal-working tools, especially the lathe and milling machine. Gunmaking involves more milling, or rotary cutting, than any other operation. The extent to which Whitney deserves credit for originating the milling machine is hotly debated. Some, claiming he did invent such a machine in 1818, place him at the top of the genealogy of American machine tool builders. Others say no, Whitney did milling but used two relatively simple methods—a circular saw or a hollow mill for turning rough forgings. The latter claim that the milling machine was invented in Connecticut in 1818 in the Middletown armory of Robert Johnson.

Before going further, let's define a machine tool and take a brief look at England's contribution. The National Machine Tool Builders' Association calls machine tools "power-operated metal working machines, not portable by hand, having one or more tools or work holding devices used for progressively removing the metal in the form of chips." The Encyclopedia Britannica adds: "These lie at the basis of all modern industrial production, for they are not only necessary for the manufacture of every class of engine and class of mechanism, but every manufactured product . . ." The master tools of industry, they are self-perpetuating. Without them the factory system could never have achieved the degree of sophistication, automation and specialization common today.

Machine tool building began in England with Watt's steam engine, which was patented in 1769 and which depended for its success upon the availability of tools that in turn could produce parts sufficiently accurate to make his engine a practical device. From the Revolution on, such giants of ingenuity as Wilkinson, Braman, Bentham, Brunel, and especially Maudslay, Nasmyth and Clement, in rapid succession invented what became the borer, lathe, planer, shaper, and steam hammer. From these sprang a whole family of machine tools for British industry. The naval architect and engineer Samuel Bentham obtained a patent in 1793 for his principle of using a rotary motion for cutting tools on both wood and metal.

A distinction must be drawn between power-driven machinery that replaced handwork and machine tools. Cotton textile machinery, both in England and in America, played a significant role in the evolution of the factory system, but its introduction alone—without machine tools—would have served only to mechanize certain basic operations, like spinning and weaving. Others would have had to be manually performed. Machine tools made possible the interchangeable system, first of all replacing wooden machine parts with iron. Frederick S. Blackall Jr. extolled them as "the touchstone which converted the age-old craft system into a modern industrial economy, brought to the common man conveniences and a standard of living which even kings had not before enjoyed, and launched a wave of creative invention and production such as the world had never known."

In the course of his struggle to defend his cotton gin patent, Whitney was in close touch with the U.S. Patent Office and may have learned of Bentham's invention. As early as 1798 he apparently had the concept for a power-driven, multitoothed cutting tool. But whether or not his was the first, in this country or abroad, he did make valuable and original contributions to the history of machines and tools. He has been particularly cited for introducing the jig (a holding device) and the gage (a measuring device). Both were undoubtedly crude, but essential to making his system work. Using a gage meant that the workman, even an unskilled one, could inspect a piece accurately and determine if it fell within the tolerance of allowable error. Although Samuel Bentham and Marc Brunel applied some of the principles of repetitive production and precision to pulley blocks for the British Navy, they overlooked the vital element of limit gages. In fact, Europe did not adopt the American gaging system until 1855. Other American gunmakers, however, were quick to follow Whitney's example. Even before the machine tool industry was established, his ideas also influenced the perfection of textile machinery in New England—until mid-century the primary heavy goods industry in the United States. Robert Fulton, who corresponded with Whitney, said of him: "Arkwright, Watt and Whitney were the three men that did most for mankind of any of their contemporaries."

A long memoir which Whitney prepared in 1812 for the War Department in support of an application for another contract summarized twelve years of his creative thinking:

> This establishment was commenced and has been carried on upon a plan which is unknown in Europe, and the great leading object of which is to substitute correct and effective operations of machinery for that skill of the artist which is acquired only by long practice and experience . . .
>
> Having actually made about 15,000 muskets, at least equal in quality to any that have been manufactured in this country (which is more than has been accomplished by any other individual in the United States) he feels himself warranted . . . in believing that the New Methods which he has invented of working metals and forming the several parts of a musket, are practically useful and highly important to his country.

Whitney left a good size fortune of over $130,000 including unpaid loans to employees and relatives, but the inheritors of this legacy did little to improve it. His only son being a mere child, he directed his nephews Eli and Philos Blake to carry on the business and complete the third contract for 15,000 muskets. They lasted only a decade, at which point the trustee of the estate, ex-Governor Henry W. Edwards, took over until Eli Jr. was ready to assume control in 1842. What he inherited could hardly have pleased him and would have outraged his father. Production had declined; the machinery had grown old, worn and even obsolete; nothing had been done to keep abreast of improvements in government arms, such as the percussion cap. Power was still being supplied by two undershot iron waterwheels fourteen feet in diameter and six feet wide, which young Whitney replaced with the new-style turbine. Possessing more than a little of his father's inventiveness, he introduced "decarbonized steel" for the barrels of the newly adopted Harper's Ferry rifle, a muzzle loader using a half-inch ball and percussion cap, and improved the machines for drilling barrels. The usually critical Ordnance Department praised the accuracy of these rifles, put them to good use in the Mexican War, and by 1856 had purchased more than 30,000. In 1858 the shirt manufacturer Oliver Winchester took over Whitneyville for making repeating arms.

* * *

The first official pistolmaker of the United States looked the commonly accepted stereotype of the Yankee. Simeon North of Berlin was on the tall side, bony and muscular, reflective and taciturn in manner, with the hands of the mechanic and the mind of an entrepreneur. There are startling similarities between North and Whitney. They were born the same year, 1765; they worked within twenty miles of one another; they shared a natural mechanical genius; at the same time they turned their attention to making firearms for the government; and both made notable contributions to production methods. Their first government contracts were signed within a nine-month period, Whitney's for 10,000 muskets on June 14, 1798, and North's for 500 horse pistols on March 9, 1799. In fact, it is a moot historical question which one initiated the interchangeable part concept, because both undoubtedly applied it almost simultaneously. Yet as far as we know, these two pioneers of the factory system were not personally acquainted.

Like his father and grandfather, North began life as a farmer in Berlin. At the age of thirty he bought an old sawmill on Spruce Brook adjoining his farm and made scythes, which he peddled around the state. When he agreed to make 500 flintlock horse pistols for $6.50 apiece, he correctly believed he was entering upon a business that would be vital to the defense of his country. He completed

Simeon North

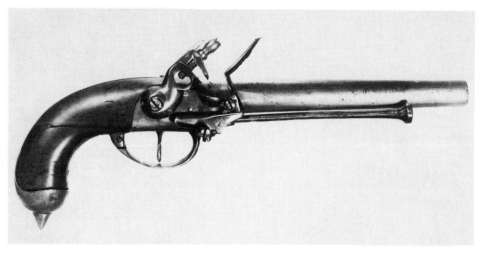

North pistol

this contract before Whitney had even made a delivery and promptly obtained another for 1,500 pistols, which included an advance payment of $2,000. For the next few years there is no record of his receiving further contracts, and he resumed scythe making. But from 1811 on he devoted his entire attention to firearms. By then he had built a large two-story factory with a forge in the basement. His workmen boarded with members of the North family, earning up to twelve dollars monthly and paying five dollars in board. Of cheerful and even indefatigable disposition, he smoked cigars but shunned alcohol, had a friendly intimacy with his nearly fifty "Armorers", as he called them, and, like Whitney, had little time for anything but work.

In common with other early manufacturers he could not find enough skilled men and realized the only practical substitute was the machine. His letter of November 7, 1808, to the Secretary of Navy reveals his understanding of the merits of standardization:

> I find that by confining a workman to one particular limb of the pistol until he has made two thousand, I save at least one quarter of his labour . . . and the work will be as much better as it is quicker made.

To carry out this division of labor, he requested and obtained an advance of $4,000 to purchase the necessary stock. In 1810, Tench Coxe, who had become the Federal Purveyor of Supplies, gave him an order for another 500 horse pistols.

Shortly after the War of 1812 started, a conflict strongly opposed by North's merchant friends in Hartford because it destroyed their maritime commerce, the Secretary of War visited his plant and urged him to expand. North then devised a new model of his pistol with interchangeable parts, took it to Washington and the next year was awarded his sixth contract, by far the largest and most important: 20,000 pistols at $10.00 per pair, to be delivered within five years, and an advance of $20,000. It came one year after Whitney's second contract for 18,000 muskets. This 1813 contract was unique in being the first in which the contractor agreed to produce arms with interchangeable parts: ". . . the component parts of pistols are to correspond so exactly that any limb or part of one Pistol may be fitted to any other Pistol of the Twenty Thousand."

It is quite possible that the North family may have been inspired by the very limited efforts of early Connecticut clockmakers like Daniel Burnap and Gideon Roberts to mass produce the wooden works of tall clocks. We know that Simeon's brother-in-law, Elisha Cheney, a clock manufacturer in East Hartford, made screws for his pistol locks. Another source of inspiration for both North and Whitney may have been the work habits of cabinetmakers like Eliphalet Chapin. It has been discovered that the components of chairs made before the Revolution can be interchanged after disassembly. It would certainly have been logical for a craftsman, when making up a set of chairs, to cut or form certain parts, like the legs, simultaneously. In turn, this practical and time-saving method may have been passed onto the gunmaker and clockmaker, both of whom would have been familiar with woodworking.

Fortified by the 1813 contract, North at once purchased fifty acres in Middletown, which offered a better source of waterpower, and erected a large dam and three-story brick plant, altogether an investment of $100,000, an unprecedented amount for those times. He turned over the management of his Berlin shop to his son Reuben, who concentrated on making forgings for the pistols. Soon he was employing seventy men and turning out thirty pistols every day. This contract, too, he completed without criticism of his quality, without a single rejection, with all advances repaid and with the rigid requirements for uniformity of parts fully met. Apparently the exclusive supplier of pistols for the United States government for the first quarter of the nineteenth century, in later years North also manufactured rifles. He outlived Whitney by twenty-seven years, and the year following his death his factory closed its doors.

Not every historian accepts the priority of Whitney's methods. There was, for example, the earlier contribution along similar lines of the mysterious Honoré LeBlanc, a gunmaker whose workshop Jefferson, while minister to France, visited in 1785. The alert Jefferson wrote John Jay that Congress might be interested in the Frenchman's method. "It consists," he said, "in the making every part of them so exactly alike, that what belongs to any one, may be used for every musket in the magazine . . .". LeBlanc almost immediately became a blank; nothing more was ever heard of him. Professor Robert S. Woodbury claims that the Springfield Armory, founded in 1794, used gages as early as 1801, while he doubts Whitney ever did. In his opinion Whitney's machines were no different from those at the two government arsenals, and he questions whether Whitney's miller took precedence, inasmuch as Harper's Ferry did milling. Contradicting this argument is the fact that Harper's Ferry did not adopt the Whitney system until 1819, while Springfield had no milling equipment until 1834.

Although North's was the first government contract to specify interchangeability, it marked not the beginning of the system but the first contractual recognition of a practice already used by Whitney. The facts seem to support Whitney's priority. While North and his three sons were applying the principles of interchangeability in 1808, we know from Jefferson's letter to Madison that Whitney, seven years earlier, had already invented molds and machines for fabricating uniform locks in his rifles. Also favoring Whitney were his national fame as the father of the cotton gin, his factory as the largest and best equipped in the country, his help in getting the two government armories on their feet, his building the first truly American machine tool, and North's relative anonymity. Moreover, Whitney, unlike North, certainly influenced other pioneers, who perpetuated, refined and adapted his ideas of uniformity and precision, especially Eli Terry on wooden clocks, Chauncey Jerome on rolled brass clocks, Samuel Colt on revolvers and Elias Howe and Isaac Singer on sewing machines. Each of them, through the miracle of low-cost production, further helped to bring the luxuries of life within reach of every man.

There is a darker side to the factory system that cannot be overlooked, one that was bound to emerge, given the social and economic implications of what North and Whitney created. It crushed the individual artisan and tended to sacrifice quality for quantity. It ended the apprentice system. It speeded up production at the expense of the working man, giving rise to problems of health and safety. It created a laboring class and the stratification of labor versus management. It attracted hordes of unskilled immigrants who had to be absorbed into American society. Yet the introduction of machine tools eliminated most of the drudgery that for centuries had been the lot of animals and men; they called for new skills and trades; they shortened the working day from twelve hours to eight or less. By producing more goods at lower cost, they created greater demand, which in turn made more jobs at higher wages. Finally, they provided the American middle class, with the highest standard of living ever enjoyed by any nation in the entire spectrum of history.

*   *   *

Despite the contributions of the Slaters, Wilkinsons and Scholfields to the textile industry, despite the revolutionary ideas of Whitney and North for gunmaking, manufacturing could hardly be called a significantly specialized occupation even by 1820. At that time 17,541 were employed in shops, while nearly three times as many (50,518) worked in agriculture. More than two-thirds of the state's 275,000 citizens lived on small-sized farms; sheep raising was popular; the annual output of hay exceeded in value all cotton and woolen goods by a million dollars. Curiously, the Yankee spirit of invention had little effect on farm tools; the cast iron plow was just coming into use; the mower or reaper did not appear until after 1850. Most of Connecticut's farmers worked their poor soil with antiquated methods.

Although peddlers had made Yankee notions famous, especially tinware and clocks, no manufacturing city as such existed. Only thirty-eight companies of all kinds were large enough to be incorporated by state charter. Seldom as much as $100,000 was invested in any one company. Only Eli Whitney and Simeon North, through their government contracts, had a national market for their products. Most manufacturing capital went into textiles. Windham County would boast twenty-two cotton mills, one-third the total in the entire state. A writer in the Norwich *Courier*, seeing in the growth of industry an end to the disturbing trend for young farmers to emigrate, commented:

> In the three eastern districts of Connecticut the traveller's eye is charmed with the view of delightful villages, suddenly rising as it were by magic, along the banks of some meandering rivulet, flourishing by the . . . protecting arm of manufactures.

There were also as many woolen as cotton mills, but they filled only local demand for cloth; their machinery, if any, was simple and crude; and they employed mere handfuls.

In 1819, Pease & Niles published their *Gazetteer of Connecticut and Rhode Island*, giving a detailed account of the enterprises in each town but grossly exaggerating their importance. Ardent partisans of manufacturing, they called every mill and shop a factory, regardless of size. The ubiquitous "shoe factories", for instance, were merely cobblers' shops, since shoemaking did not become industrialized until after the invention of the sewing machine in 1846. Danbury's fifty or more hatters were likewise tiny proprietorships or partnerships. Mix, Barney & Company, buttonmakers employing twelve workmen who produced up to $500 a month, was considered a large firm. Most articles of necessity were still homemade. Nearly every family wove its own cloth and took it to the fulling mill for finishing. The local blacksmith was especially versatile, able to fashion, like James North of New Britain, an amazing variety of tools from augers to wedges. For most villagers, manufactured goods were rare luxuries reserved for the well-to-do. Household industry still prevailed.

While New London specialized in fishing, Norwich had six mills. New Haven turned out hats, nails, powder, combs and shoes, besides having eight carriage shops. In 1820, Whitneyville employed fifty-three men and boys but for the past decade had yielded less than five percent profit on the capital invested. Humphreysville, since the death of its owner the previous year, was all but idle. Middletown was known for its pistols, swords, buttons, pewter, tinware and combs.

Largest of all with a population of just under 7,000, Hartford displayed the most industrial activity. In addition to fifty small shops and mills, the town had six churches, twelve schools, twenty-one taverns and—to quench a sudden thirst—eighteen ale houses. Blacksmiths, masons and cabinetmakers were plentiful. The *Gazetteer* listed such businesses as:

> 1 Machine card factory, which manufactures $10,000 worth of cards annually. 1 Whip-lash factory, which manufactures $10,000 worth of the article annually. 2 Hat factories, one of which is upon an extensive scale, and employs 36 workmen. 2 Looking glass factories, which together manufacture $30,000 worth of goods annually. 4 Coppersmiths, two of which carry on the business upon a large scale; one of them employing about 20 workmen.

There were also a cotton mill, two woolen mills, an oil mill, six tanneries, five potteries, one buttonmaker, two tin shops, one maker of Britannia ware, one bell foundry, fifteen shoe shops, six book binderies, three lottery offices and eight distilleries.

State-wide, Pease & Niles reported manufactures other than textiles as follows:

Grain mills	—	513
Fulling mills	—	225
Wool carding mills	—	235
Forges & furnaces	—	64 (44 in Litchfield County)
Paper mills	—	24
Oil mills	—	17
Powder mills	—	13 (11 in Hartford County)
Anchor shops	—	8 (all in Litchfield County)
Glass works	—	5
Slitting mills	—	3
Gun factory	—	1 (Whitneyville)

Mention was also made of buttons, spoons, ivory combs, plows, tinware, clocks, hats and leather goods, silk and shad fishing. With the exception of Eli Whitney's armory, however, none could properly claim the name of factory.

Once launched, as it clearly was by 1820, the manufacturing movement gained a momentum that could not be stopped. At the first national fair of

Hartford Iron Foundry, 1821, predecessor of Woodruff & Beach

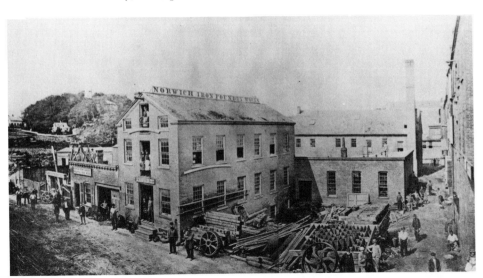

Norwich Iron Foundry

industry held in Washington in 1825, visitors were astonished at the scope of American products developed mainly in one generation. During the 1830s textile mills, both cotton and wool, outstripped family spinning and weaving, finally ending the era of household industry. To run the noisy spindles and looms, steam and coal rapidly replaced water and wood as sources of power. American mill hands now outproduced their counterparts in England, but most goods were of coarse quality like blue jean and muslin. This was the decade of fastest growth for manufacturing of all kinds in both New England and the Middle Atlantic States. Until the Civil War manufacturing thrived, except for two financial panics, wholly in an environment of encouragement, peace and tranquillity.

CHAPTER IV

Peddlers Afoot and Afloat

Connecticut . . . owes (her) reputation to the "pedlars" of "Yankee notions" whom she sent forth, especially into the South—a tatterdemalion crew of hardbitten young men who straggled down every road and bypath of the countryside, pinching pennies, higgling and haggling, flying by night,. and leaving behind them, according to popular belief, a trail of wooden nutmegs and of clocks that would not go.

ODELL SHEPARD

THUS did Odell Shepard characterize Connecticut's contributors to the art of salesmanship and to the creation of a widespread market for its goods. Peddling, of course, is one of the world's oldest professions. But the Yankee peddlers were a special breed, carrying wares on their back, in a saddlebag, cart or the hold of a small ship, wandering far and wide, usually alone. Bartering or selling the plain and exotic wants of man, the peddler performed an essential service: the distribution of products. Without him trade beyond the nearest towns would have been impossible. Invariably lean, dirty and unshaven, both loved and damned, welcomed and shunned, he was also a purveyor of news, gossip and fashion, sometimes a charlatan or even a spy. America's first travelling salesman had, as Walter Hard said, "a unique character and his business was a unique institution."

Around him and his jingling cart, really a mobile department store, hung an aura of excitement and romance. His noisy arrival in the autumn along muddy, dusty roads leading to our frontiers signalled a special occasion, a kind of one-man circus and fair. From the Revolution to the Civil War, from Berlin to Savannah in the South, Detroit in the West or Quebec in the North, his was a familiar figure. His glib tongue, wit, cunning, impudence and coolness earned

A peddler's wagon

him the caustic sobriquet "Sam Slick". A British traveler learned that the Southern businessman regarded his Yankee counterpart as "a commercial Scythian, a Tartar of the North whose sole business in life is to make inroads on his peace and profit . . . There is no getting rid of them . . . he has the grip of a crab, with the suction of a mosquito; you can't deny, you can't insult, you can't fatigue him; you can only dismiss him with a purchase . . ." The peddler's honesty and sobriety were always subject to question:

> The manner in which this ware is disposed of puts to flight all calculation . . . A multitude of these young men direct themselves to the Southern States . . . Each of them walks, and rides, alternately, through this vast distance . . . He wanders into the interiour country; calls at every door on his way; and with an address, and pertinacity, not easily resisted, compels no small number of inhabitants to buy . . . This business is said to yield both the owner and his agent valuable returns . . . No course of life tends more rapidly, or more effectually to eradicate every moral feeling.

So wrote Timothy Dwight early in the nineteenth century. It is true that the fast-talking peddler often resorted to flimflam. He might on occasion be guilty of selling wooden hams, cheeses and nutmegs, or clocks that didn't run, but shrewdness in bargaining has always been an accepted part of salesmanship. By the rather low standards then in effect, he was not immoral—only a consummate liar.

Timothy Dwight also claimed that peddling as an occupation began with two enterprising Scotch-Irish immigrants, who must have been among the very first from that country to settle in New England. In 1740 William and Edward Pattison chose to make their home in the village of Berlin, not far removed from the Connecticut River. Finding the land unsuitable for farming, they set themselves up to manufacture tinware, a trade they had learned in England and one requiring little capital and no waterpower. The brothers imported sheet tin from England, carried it from the port of Boston on horseback to Berlin, pounded it into shape with wooden mallets over anvils, and then polished the finished product until it shone like silver. Their shop was such a noisy place the neighbors dubbed it "Bang-all". When enough tinware had been made to fill a sack or two, the Pattisons walked to nearby settlements in the area of Hartford, Middletown and New Britain, calling on farmhouse after farmhouse until their stock was exhausted. If the farmer had no cash, they took goods in trade. Soon they were using horses equipped with baskets, then training apprentices who transported their wares further afield after the Revolution—teapots, lanterns, candlesticks, plates, cups and pans. With the advent of turnpikes, they designed a special wagon containing a maze of hooks, drawers and secret compartments that would hold $600 worth of merchandise. The first red or yellow-painted wagons were drawn by one horse, later by two or more. At the end of his peregrinations, lasting six to eight months and often covering more than a thousand miles, the peddler would usually sell his horse and wagon and return to Connecticut.

The Pattisons' neighbors were quick to copy their techniques and to make Connecticut the undisputed leader in the tin industry down to 1850. Little "bang-alls" appeared in Bristol, Cheshire, Meriden and Wallingford. In Berlin alone, by 1815, 10,000 boxes of tinned plate were being hammered annually into what Dwight called "culinary vessels". In the typical shop, employing twenty to thirty persons, the workmen frequently went forth in the fall to sell articles they made during the summer. As methods improved and sales picked up, five tinworkers could turn out enough stock to keep twenty-five hawkers on the road almost full-time. Rather than independent tradesmen, they were manufacturers' representatives backed up by supply stations located at strategic points. According to one historian, "the most valuable contribution which tinware manufacture made to the industrial history of the state was the trade organization which it perfected." Selling the product was far more demanding than its manufacture.

Later on, during the 1820s, itinerant vendors carried an astounding variety of small articles, or specialized in one item like clocks, chairs, books or buttons.

In addition to tinware, the peddler's load might include pins, needles, scissors, combs, children's books, cotton stuffs, hats and shoes, paper, axes, saddlery and various dry goods. Such a wagonful represented an investment of up to $2,000 for the proprietor; the peddler himself might earn as high as $40.00 a month plus a half share of the profits. E. C. Maltby of Northford kept four carts on the road filled only with buttons of bone, ivory, horn, and wood. Middletown contributed "gum elastic" suspenders or galluses; East Hampton added bells; Windham County, thread. Hartford around 1810 depended on peddlers for the distribution of subscription books like *The Cottage Bible* and Webster's *Dictionary*, which helped to make the capital a major publishing center. The peddler even had such frivolous items as the Jew's harp which, along with the trumpet and drum, were the only musical instruments sanctioned by the Congregational Church.

Not only did their nasal cry become well-known, but the Yankee peddlers succeeded in popularizing a great many Connecticut-made products and encouraging a host of new shops, all eager to reap the harvest of peddled wares. In 1793 the clock pioneer Eli Terry began his illustrious career by selling wooden clocks on horseback. Chauncey Jerome did likewise twenty-five years later. Ivory combs were supplied by the Pratts of Essex, buttons came from the first brass mills in the Naugatuck Valley, Britannia ware and coffee mills from Meriden. Ed Jenkins of Colebrook, who made the first "elastic steel" fishhooks in 1813, owed his success to peddlers. Even the first silver-plated spoons turned out by William B. Cowles and Asa Rogers in 1843 at their Spoonville shop found no market until the hawkers, who were already handling Meriden table cutlery, took out samples and found ready buyers. At Riverton, Lambert Hitchcock shipped his chairs "knocked down" to be assembled by Yankee peddlers.

When house-to-house selling dropped off around mid-century, Thomas Danforth of Rocky Hill, an early pewterer and outfitter of peddlers, established the first chain-store business in America by having his sons manage retail outlets in Philadelphia, Atlanta and Savannah. A latter-day peddler saved the inventor of the lathe chuck, Simon Fairman, from bankruptcy just before the Civil War. Some peddlers became famous: Benedict Arnold of Norwich as a young man sold woolen goods in northern New York and Canada; P. T. Barnum, the Bethel practical joker and pitchman, peddled entertainment. Collis P. Huntington of Harwinton, founder of the Southern Pacific Railroad, began to support himself at the tender age of fourteen by working on a farm for seven dollars a month plus board. When he had saved $175, he entered the watch business in New York. For several years before reaching his majority, he peddled watches and collected overdue notes throughout the South; then he graduated to the merchant class and, in 1849, joined the gold rush to California.

One peddler who achieved fame outside the business world was Amos

Bronson Alcott, the Transcendentalist philosopher from Wolcott, who preferred carrying a pack to getting an education at Yale. Alcott's five forays into the South from 1818 to 1823 proved that not all peddlers prospered, financially speaking at least. Looking back on his experiences in later years, he often said they were far more enriching than college could have been. Accompanying his cousin on a trading venture by foot into western Massachusetts, when he was only fifteen, gave him a yen for seeing new places and meeting new people. In October, 1818, along with a number of other peddlers and tinmen, he sailed from New Haven on the sloop *Three Sisters*. At Norfolk he bought two handsome japanned tin trunks and filled them with $300 worth of stock: "combs of tortoise shell of the latest fashion, jewelry & amulets & garnets & pearls, reticule-clasps & rouge-papers, essences & oils & fine soaps & pomatums, silver thimbles & gold & silver spectacles with shagreen cases for all ages, sewing silks & cottons & threads & buttons & needles with silver & gold eyes, pencil cases, pen knives, scissors of Rogers' make, with steel purses, playing cards, & wafers. And, for gentlemen, genuine magnum bonum rasors & straps. Also picture-bricks and puzzles for children. Then we have fans & fiddle-strings, etc. etc." Soon he ventured forth among the plantations on the James River and Chesapeake Bay, managing to make a dollar a day. He slept on the ground or in slaves' quarters. By April, having cleared a little over $100, he returned home.

The following November Alcott set out a second time, taking with him his younger brother. They brought back $165 to their aging father. His village of Wolcott now sent forth ten or fifteen young men like Alcott every year. His third journey in 1820 with his cousin was fraught with danger, privation and financial loss. The next year, with money borrowed from his father, he bought a wagon in Berlin, filled it with goods obtained on credit in Meriden and began the long ride to Virginia with his brother Chatfield and cousin Thomas. This too was a complete failure. He had to walk the 500 miles back, passing through New York barefooted. He arrived at his Spindle Hill farm in July with sixpence in his pocket, "many penitences at heart" and owing his father some $600. Nevertheless, he resolved to try once more, but this turned out to be the most disastrous of his trips, leaving him worn and dispirited. "From that time forth," comments his biographer, "he was to peddle only immaterial wares."

In his autobiography Governor Wilbur Cross vividly recounted the peddler's contribution to the Mansfield silk industry and the success story of a relative:

> There was a brisk demand for sewing silk manufactured in the Gurleyville district, which was run off on spools by pretty girls who easily found husbands. These girls were nicknamed "spoolers". Down to the time of the Civil War and somewhat later sewing silk from the Gurleyville mills was distributed by local

pedlars, many of whom were young men who wanted to see the world outside of Mansfield as well as to make a little money. I can imagine them as they set out on foot, with flowered carpetbags filled with silk, one in each hand, for neighboring towns within the State or across the borders. Their customers were housewives and small country stores . . . So easy was it to sell silk thread that a young man who failed to make good was called a "good-for-nothing" for the rest of his life. One such fellow came back from a fortnight's trip with his carpetbags as well stuffed as when he started out. "What," his father asked, "have you got in them bags?" "Silk," was the reply. "Didn't you sell any of that silk?" "No," replied John. "Were there no inquries?" "One man," replied John, "asked me what I had got in them bags, and I told him it was none of his damn business." Everybody laughed whenever that story was told.

As the manufacturing of silk grew, the product was distributed in larger quantities by pedlars with horse and wagon, who drove north to the Canadian border and south as far as Georgia. Of these pedlars on a large scale, none was more successful than Ebenezer Gurley, a cousin of my grandfather's, whose shrewdness led him on to a fortune. Out of his profits as a pedlar, he accumulated enough funds to become a middleman between the importer of raw silk from the Orient and the manufacturer. On one occasion he was able, with the assistance of a New York importer, to get control of all the raw silk on the market and all that was on the ocean due to arrive in port. By this corner of the market he made a comfortable fortune, and settled in Mansfield as a farmer on spacious lands by the graveyard of his Scottish ancestors, all of whom he had outstripped in the virtues of his race. Ten years after Ebenezer Gurley's clever stroke, the speculator and pedlar . . . had disappeared.

Contrary to Dwight's claim, the Pattisons do not deserve credit for being the first peddlers. Before their arrival the colony encouraged Connecticut vendors to ply their trade, paying twenty shillings for every hundred pounds worth of goods they carried. By 1727, however, local merchants were petitioning the General Court to suppress the "Multitudes of foreign or Peregrine Peddlers who flock into this Colony and travel up and Down in it with Packs of Goods to Sell". Apparently, a native-born hawker could be tolerated but not an outlander. In 1765, the General Assembly raised the cost of a peddler's license, from five to twenty pounds. The licensing system for "pedlars, hawkers and petty chapmen" was later abandoned on condition that no foreign products be sold, but the wheeling and dealing peddler continued to be an irritant to many merchants and even some manufacturers. In 1828, Calvin Butler of Plymouth complained the unlicensed peddler was a menace to industry because he could set his own prices, and he asked for the restoration of licenses. The legislature rejected his bill and even removed the ban on selling imports, because of the impossibility of distinguishing them from domestic merchandise. But in 1841 it reestablished

Hartford, 1840

Shipbuilding on the river

City of Hartford

Rocky Hill steam ferry

a vendor fee of twenty dollars and relieved the retailer of unfair competition.

Regardless of his origin, the Yankee peddler became "the whipping boy of our early national life, the butt of national jokes and the target of all visitors to America." How many Connecticut sent forth every year no one knows, but they must have numbered well into the hundreds. In 1829, Meriden counted 40 and Hartford claimed 60 more. In a single year 147 patronized the Wadsworth Tavern in Hartford. Some taverns refused admission to "tinkers". They were immortalized in early fiction: James Fenimore Cooper's spy was one; Rip Van Winkle's wife died from a fit of anger over a peddler; the first humorous folk figure, Sam Slick, wandered through New England and Canada disposing of clocks at inflated prices to reluctant but outwitted buyers. Later, he was pictured as a Jewish notion hawker, since German Jews began to replace Yankees after 1835. His decline began with the coming of the railroad in the 1840s, and soon he was transformed into a country storekeeper or a "drummer" who rode the rails for his company, although in the Fuller Brush man he had a twentieth-century heir. His place in early industrial history will always be assured, however, not for his flamboyant character, but for bringing together manufacturer and consumer during the period when transportation and communication were in their infancy, and in giving Connecticut industry a headstart in the race to gain a national market.

* * *

Not all peddlers travelled by land. Many, as captains and merchants, used the Connecticut River, the Housatonic, the Thames and the waters beyond as a highway for trade with Boston, New York and southern ports, with the West Indies and on over the horizon. For 200 years those who sought their personal calling afloat depended upon the river and sea ports. The resulting maritime trade, as has been noted, led to a merchant-capitalist class who eventually turned to manufacturing.

The Connecticut River in particular has long been a valuable natural resource for commerce. Called by Indians "the long tidal river", it is the only body of water running the full length of New England, some 400 miles from the Canadian border to Long Island Sound. Originally the hunting and fishing grounds of the peaceful Algonquins, then a trading post for the enterprising Dutch, finally settled by the land-hungry English, the river valley was the birthplace of such pioneering achievements as the first ferry and canal, the first U.S. war vessel and the first self-government. Most occurred along the sixty miles within the colony of Connecticut, the water of which Timothy Dwight boasted was "everywhere pure, potable, perfectly salubrious, and inferior to none in the world for the use of seamen in long voyages."

Oliver Ellsworth

As far north as Hartford the Connecticut River is tidal and navigable. The sandbar at its exit into the Sound, formed by the conflux of river and tidal currents, precluded any great port rising at its mouth. From earliest times tiny river-built sloops made regular trips to the Caribbean, livestock on deck; cruised from island to island, bartering farm produce; and then returned to river wharves, where they became floating stores. The voyage upstream frequently took two weeks, as long as the sail from the land of molasses, rum and sugar to Old Saybrook. Rum was by far the leading import. Tippling, in fact, was so prevalent that an early almanac contained this ditty:

> Ill husbands now in taverns sit
> And spend more money than they git.
> Calling for drink, and drinking greedy
> Tho many of them poor and needy—

Illustrative of the wanderlust of these seafaring Yankees and of their thirst for trade was the voyage of the *Neptune*. From East Haddam, in command of

Captain Townsend, she set sail in 1787 for China with a general cargo and $500 in gold coins for emergency use. In the West Indies Townsend traded a portion of his cargo for rum and sugar. Then he sailed on to Rio de Janeiro to barter for Brazilian products. Cruising southward, he caught seals in the Falkland Islands. Around Cape Horn he continued to the South Sea Islands, where he carried on a brisk trade, exchanging calico, cotton cloth and brass wire for pearls, divi-divi, spices and dragons' blood. At last he hailed the China coast and traded his sealskins for tea, silk, ivory, lacquer work and sandalwood. The first Yankee ship to reach the Orient and circumnavigate the world, she arrived home thirty months later with a cargo valued at $280,000 and her gold unspent.

By 1750, river shipbuilding had reached its peak and continued for another century almost without interruption, despite two wars and the introduction of competing forms of transportation. River yards became famous in the East for launching ships much larger than commonly sailed out of river ports, even three-masted barks of such deep draft that they could be towed down only on a high spring current. Middle Haddam alone was reputed to have built more than 200 deep-water vessels, many for the trans-Atlantic mail and passenger service.

After the Revolution river commerce revived for several decades. With one merchant for about every fifty residents, there was no doubt of Hartford's dominating interest. Typical advertisements in the *Courant* read: "West Indian goods as usual, and a small assortment of English dry goods at wholesale," or "2,000 bushels best Turks Island salt." Most merchants owned their own ships, as well as wharves and warehouses. Nearly 200 vessels sailed regularly along the coast and southward, carrying fish, lumber, corn, grain, dairy products and ropes of red onions. Some brought back impressive earnings. For instance, the *Lorena*, 527 tons, out of Essex, cost $28,000 to build and returned a profit of over $100,000 in seventeen years.

One thriving industry along the river was the quarrying of Portland sandstone. As early as 1665 the soft, reddish-brown stone was discovered to be ideal for building purposes. There were three quarry firms active at one time. The first, Shaler & Hall, took over the oldest site from Middletown in 1788 and three years later moved to a new location further back from the river. They employed about thirty men to cut and shape the stones, which were then dragged down to the river's edge by oxen or horses and shipped on especially-built schooners. The brownstone houses in New York owe their faces to Portland, as do many of the old homes, monuments and grave markers in the Valley.

Above the original site the Middlesex Quarry Company began operations in 1819. Erastus and Silas Brainerd from Middle Haddam started a similar business, which lasted from 1812 to 1884. At its peak they had a force of 300 men and 16

Poling a barge upriver through the Windsor Locks Canal

Barnet

stone carriers. Steam derricks were used to hoist the stone from the pit, which was excavated to a depth of two hundred feet. By mid-century, some 900 men worked in the three quarries.

By 1807 Connecticut merchants met with hard times. Money became tight, exports suitable for the European market scarce, and the British closed the West Indies to American ships. Wadsworth found relief by importing tea from China and shipping ginseng root from the Valley in return. Both the Embargo on exports and the War of 1812 with England brought about a coastal blockade that caused severe distress. During the latter trouble, English men-of-war boldly invaded the river, set fire to Essex, and burned twenty-three ships, while American soldiers hid in a nearby tavern. To reduce their risks, Hartford merchants entered into partnerships, taking shares in various vessels and adventures. There was even joint underwriting of ship insurance, at five and one-half or six percent interest, with individual liability commonly limited to 100 pounds. From these experiences emerged the concept of financial protection against the hazards of life, or insurance, which later became one of the chief callings of Hartford Yankees.

A new era of power that had a profound impact on river commerce and later on industry began in 1815, when the steamboat *Fulton* arrived in New Haven. Rigged as a sloop, in case sails were needed—as indeed they often were at first—she made a dreadful din between the thrash of her paddles and the sharp, staccato blasts of her wood-fired engine, as steam escaped from the cylinder. Her machinery was described as "a little less . . . than is now put in a cotton mill," yet her size was imposing: 135 feet long, 327 tons. Sixty passengers made the trip from New York in about ten hours. The fare: $6.00. Subsequently, she churned up the Connecticut River, villagers rushing to the banks to see the new-fangled contraption "coming on wheels in the water," and docked in Hartford. Enthused the *Courant*: "Indeed it is hardly possible to conceive that anything of its kind can exceed her, in elegance and convenience." She was the last to be designed by Robert Fulton, generally acknowledged as the inventor of the steamboat, despite the fact that Connecticut's John Fitch applied steam successfully to the propulsion of vessels through water twenty-two years before Fulton's *Claremont*.

Three years after the *Fulton*'s debut a steamboat launched at Hartford served as a towboat along the river. There quickly followed regular service thrice weekly on the *Enterprise* of Captain James Pitkin, which advertised: "Passengers can be landed at any place on the river at their pleasure." The 112-foot-long *Oliver Ellsworth*, in 1824, was the first of a long line of floating palaces over the next half century. The same year, "amidst the salute of cannon and the shouts of thousands of gratified and grateful spectators," the Marquis de Lafayette, after revisiting Hartford, left aboard the *Oliver Ellsworth*.

With their crude crosshead engines and undependable copper boilers, travel on the early sidewheelers was at best a hazardous undertaking. Steaming one evening in the Sound, four miles from Old Saybrook, the *Oliver Ellsworth*'s boiler exploded, injuring several persons and killing a fireman. She managed to sail to the town dock, from where an excited post rider galloped to Hartford, burst into the State House, where the legislature happened to be sitting, and shouted: "the Eliver Ollsworth biled her buster!" Not long after, the *New England* blew up at Essex, killing or maiming fifteen out of seventy people aboard. Notwithstanding such disasters, after 1840 the number of steamboats increased sharply.

After 1825, when the U.S. Supreme Court declared unconstitutional Robert Fulton's and Robert Livingston's steamboat monopoly, a number of companies competed on Long Island Sound. The vicious commercial warfare attracted opportunists like the hard-driving Cornelius Vanderbilt, who cut rates and schedules and launched larger, more sumptuous vessels. His *Water Witch* boasted of making the New York run in only thirteen hours. Ticket prices fell from five dollars to as little as one dollar plus meals. River traffic was at its peak, though the railroad whistle would soon sound its deathkneel. In one year alone Hartford reported over 2,000 arrivals and departures at her twenty or so wharves, even though her population barely topped 13,000. One writer called the Connecticut River the Mississippi of New England. The regular trade route extended from the river ports to New London, where the shallow draft sloops outbound for the West Indies filled their holds. On their return, they stopped again at New London to unload cargo. Over half of the Connecticut Valley trade was with New York, mostly in granite, brownstone, food products, cider and gin, wood and fish.

Above Hartford, when the steamboat arrived, the Enfield rapids still limited navigation to small flatboats able to transport over them less than ten tons by poling and sailing against the current. It was common to see rafts floating downstream laden with hides, lumber and stone. The flatboat, some seventy feet in length, usually had a cabin in the stern in which the captain and crew lived; amidships was a mast with a large squaresail and topsail. Passage was possible only in the daytime. At night they tied up along the banks. Taverns near the shore supplied crews with meals and lodgings for twenty-five cents each. River men were merry, heavy-drinking stalwarts who, to make sure of their daily ration, found rum an excellent cure for blistered hands.

Hartford merchants were frustrated, however, by their inability to make full use of the river for trade with Massachusetts, New Hampshire and Vermont. In 1822, their business was threatened by the granting of a charter for constructing the Farmington Canal from New Haven to Northampton, where it would join the Connecticut River. Improved roads and the stagecoach were also competitors.

Meeting in Joseph Morgan's coffeehouse, the best-known hostelry of the time—a club, chamber of commerce and tavern rolled into one—the merchants formed the Connecticut River Company and obtained a charter for improving upstream navigation. Believing the river far superior to any land canal for economical transportation, they immediately decided to prove it with a steamboat. In mid-November, 1826, a sternwheeler called the *Barnet*, only seventy-five feet long and drawing less than two feet, arrived at Hartford, having been towed from New York. She caused a tremendous sensation. One man who followed her some distance along the shore marvelled that the boat went just as fast as he could walk. With difficulty she mounted the falls. At Springfield "twice 24 guns announced and welcomed her arrival . . ." The news flashed around the countryside:

> This is the day that Captain Nutt
> Sailed up the fair Connecticut.

At every stop until she made Brattleboro, Vermont, two weeks later, she was met with cheers and the firing of cannon. Most country folk had never seen a steamboat before. At Bellows Falls a banquet was tendered the crew and effusive toasts drunk: "Connecticut River—Destined yet to be the patroness of enterprise, and to bear upon her bosom the golden fleece of industry," and "the grand highway from Canada to the seaboard. Give us steam!" Her backers were ecstatic. The ascension was celebrated by a great dinner at Morgan's coffeehouse. But although the *Barnet* proved the ability of a steamboat to go upriver at six knots against current and wind, she met an untimely end. A year later the company directors voted to pay the "funeral charges of Mr. Joseph Groumly who was unfortunately scalded to death by the bursting of the boiler."

Encouraged nonetheless, the merchants proceeded with the building of the Enfield Canal, a six-mile long, seventy-foot wide ditch that would be deep enough to accommodate large flatboats and steamboats up to seventy-five tons. Four hundred Irishmen were imported as workmen, their worldly goods tied in red bandanas, and in 1829 the canal opened to traffic. Fifteen boats passed through the first day. Soon sternwheelers were chugging daily between Hartford and Springfield. Tolls were one dollar per passenger and fifty cents per ton of freight. In 1842, the novelist Charles Dickens, on a visit to America, made a downstream trip in the *Massachusetts*, the engine of which he described as having about "half a pony power". Actually, it was rated at nearly twenty horsepower.

The success of the Erie Canal had inspired the digging of an overland waterway to connect New Haven with Northampton, a distance of eighty miles. Eventually, it was supposed to reach the St. Lawrence River. Chief engineer for the ambitious project was Benjamin Wright of Wethersfield, designer of the Erie

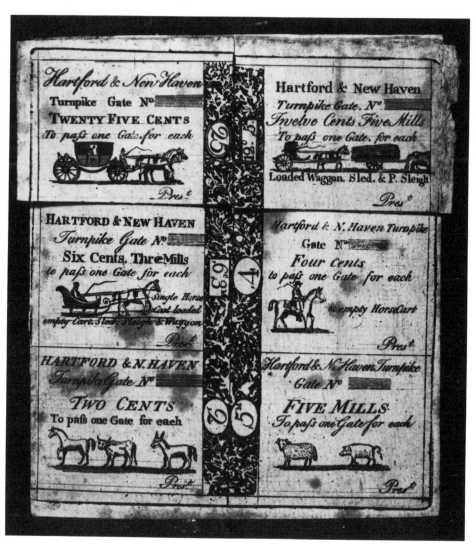

Turnpike tickets

Canal. Seventeen towns participated in the Farmington Canal, but from the very beginning it was doomed. Governor Oliver Wolcott Jr. broke the spade he wielded at the groundbreaking, July 4, 1825. Floods, washouts, droughts and debts plagued the investors during the canal's twenty-year life. Once, to settle an old grievance, a farmer caused a major break at Cheshire by cutting the bank to flood his neighbor's property. Various towns used its convenient water to put out fires. To raise more capital, the towns persuaded the state to withhold charters for two new banks in New Haven until they promised to subscribe for large amounts of stock. In 1835, the canal was completed as far as Northampton; packets made the trip in a single day, charging the modest sum of $3.75 including meals. Its backers in the end lost more than a million dollars. But the Enfield and Farmington canals, as well as turnpikes and steamboats, were doomed anyway by the coming of the railroad.

In building railroads Connecticut lagged five years behind Massachusetts, mainly because its waterways, plus an elaborate system of turnpikes, took care of transportation needs. Moreover, canal, steamboat and turnpike interests bitterly opposed this upstart form of competition. Belatedly, in 1838, a legislative committee called attention to its advantages. Railroads were seen as the means of stimulating employment, halting emigration, reducing the cost of moving manufactured goods and increasing the state's taxable property. The committee advocated state loans, as Massachusetts had granted, but the legislature preferred to let private enterprise carry the full load. The first rail service was inaugurated in November, 1837, over a five-mile track from Stonington east to Rhode Island, where a line had been built to Providence. On this occasion Governor Edwards rode in the "superb cars, handsomely decorated with miniature American flags," and with 400 other guests feasted on partridge, turkey and champagne at Stonington. The first locomotives, wood-fired, weighed about six tons; the coaches were nothing more than stages—with four wheels, leather springs and side doors. Eight passengers could squeeze inside, and an equal number piled onto the roof. Guards sat in front and back. Later the noisy and dust-filled coaches were enlarged to accommodate up to sixty people, with such amenities as tallow candles and stoves inside.

For the Hartford to New Haven line, which received a charter in 1833, capitalists in New York and Connecticut put up almost $1,500,000. When the panic of 1837 slowed stock subscriptions, the chief financial burden fell upon James Brewster, a wealthy New Haven carriage manufacturer, who retired from business to raise the rest of the funds. Helped by short-term loans at high interest rates from Hartford banks, the route was finished in 1839. Irish hands, who were paid all of seventy cents for a ten-hour day, laid the tracks. To prevent through travel to New York, the Connecticut Steamboat Company bought up all the

Hartford Railroad Station

Portland stone quarries

steamboat connections from New Haven; the railroad countered by getting permission from the state to have its own vessels. Five years later the line reached Springfield. At the same time the legislature authorized its extension to New York, and it was completed in January, 1849.

Capital for this last link in a continuous route that now connected Maine in the north to Georgia in the south came from New York. One backer was Anson G. Phelps, a New York merchant and founder of Ansonia. A stock fraud and wreck nearly ruined the company a few years later. Robert Schuyler's manipulations drained the railroad as severely as the $300,000 in damages that had to be paid when a train ran through the open draw of the bridge across the Norwalk River, in 1853, killing forty-five passengers.

By 1850, fifteen railroad companies had been organized. Not until the Civil War, however, did they become really profitable or benefit manufacturing. From then on they proved a great boon, accelerating the growth of urban centers and simultaneously assuring the stagnation of those rural towns not fortunate enough to have access to the iron horse.

CHAPTER V

The Soulless Corporation

WERE it not for that curious creature of the law called a corporation, business as we know it today simply could not exist. Paralleling the rise of the factory system in Connecticut during the early nineteenth century were changes in company organization and structure in response to the financial exigencies which beset the early entrepreneurs. From proprietorship to partnership to corporation became the typical growth pattern. The manufacturing corporation was nowhere pursued with greater vigor and success than in Connecticut, which contributed to America the first general law of incorporation. More than any other legislation, it stamped the word "free" on the American enterprise system and accelerated its development in a democratic way.

A favorite aphorism in the time of Andrew Jackson held that "corporations have neither bodies to be kicked, nor souls to be damned." For a long while they were regarded with suspicion and distrust as instrumentalities of special privilege, monopoly and economic despotism. Chief Justice John Marshall defined the corporation as "an artificial being, invisible, intangible and existing only in contemplation of law." Its unique characteristics, he added, are "immortality and individuality, properties by which a perpetual succession of many persons are considered as the same, and may act as a single individual. They enable a corporation to manage its own affairs, and to hold property without the perplexing intricacies, the hazardous and endless necessity of perpetual convey- ances for the purpose of transmitting it from hand to hand." No better means—none more efficient, flexible nor lasting—has ever been found to utilize the ambitions, the energies, the creations of free men in their pursuit of wealth and power.

Although corporate enterprise had its origin in England's sixteenth-century trading companies, Marshall's definition did not apply to their loose federations. Members of the East India, London and Plymouth companies paid their own debts and assumed full responsibility for their own liabilities. Mainly concerned with extending the British Dominion, they were franchises given by Parliament with royal approval. In colonial days the English form of incorporation was copied by those businesses operating in the public interest, such as banks and turnpike companies, each of which had to seek approval from the Court of Common Council and, later, the General Assembly.

The first American corporation to obtain a charter, in 1709, was for mining copper in East Granby, then part of Simsbury. The discovery, four years earlier, of a vein of copper ore in the yellow sandstone of Talcott Mountain suddenly made the village of Simsbury the talk of New England. When the good news was revealed in town meeting, the excited voters at once made sure their rights were protected by "reserving for ever for the Towne use and disposall all such mines or mineralls." The next year sixty-four of the voters became shareholders in the venture, representing most of the property owners. In the contract with the "undertakers", who agreed to smelt and refine the copper, the townspeople, in a burst of charity, provided first for educational needs: ". . . off Every parcell of Coper thus refined or wrought, before any division be made thereof, the tenth part of it . . . shall be taken from it: for pious uses: (viz) two thirds of it shall be to the maintaining an able Schoolmaster in Simsbury the other third part shall be given to the use of the Collegiate school errected within this Collony. . . ." The college referred to, of course, was Yale. One "undertaker" was the Rev. Dudley Woodbridge, the local minister who, being the most learned man in town, seemed best equipped to fathom the mysteries of ore refinement. In addition, the town levied a tax of ten shillings per ton. In 1721, smelting works were set up on Hop Brook and skilled Germans from Hanover imported as workers—all in great secrecy because the English Parliament frowned on any domestic manufacturing that might compete with the mother country.

Ore from the Newgate mines yielded about fifteen per cent of vitreous copper in every ton. Before the Revolution a succession of companies sprang up—in Hartford, New York, Boston, London and even in Sweden—to finance the mining operations. Some paid fancy prices for leases. Altogether, close to a million dollars was sunk into the hills of East Granby, an astounding sum for a colonial enterprise. The poor grade of ore, the difficulty of extraction, the risk of shipping to England what was not refined, these were serious enough drawbacks to result in most investors never seeing their money again. From 1773 to 1827 Newgate served as the ill-famed state prison, at first accommodating Tories and prisoners of war during the Revolution. Its underground dungeons were described as "hell on earth." In 1836, the property was sold to the Phoenix Mining Company, which for a short period raised large quantities of ore for shipment to England. Its superintendent, Richard Bacon, saw a future for himself not in mining, but in making mines safe. Through his vision the safety fuse was

Newgate

First silk mill at Hanks Hill, Mansfield, 1810

brought to America and to Simsbury almost as soon as it appeared in England.

In 1732, the New London Society United for Trade & Commerce, a land bank, became the second American corporation. Because it made its own currency, the legislature rescinded its charter in less than a year. Next in order was the Union Wharf Company of New Haven, 1760.

The little town of Mansfield, however, holds the distinction of being the home of the very first *manufacturing* corporation in the United States. Here in January, 1789, when the population numbered around 2,600 (growing no larger until after 1920), was founded "The Directors, Inspectors and Company of the Connecticut Society of Silk Manufacturers." Five years earlier the state had voted premiums for growing silk, with which the South had experimented from the time of the Jamestown Colony and which thrived for a while, especially in Georgia. Connecticut offered ten shillings per annum for three years to whoever planted 100 mulberry trees and three pence for every ounce of raw silk. Dr. Nathan Aspinwall, an agricultural enthusiast who had a mulberry orchard on Long Island but lived in New Haven, was the chief promoter of silk culture, along with President Ezra Stiles of Yale. He set out an orchard in Mansfield in 1760 and two years later introduced silkworm eggs. Although every parish received a half ounce of mulberry seed from Dr. Stiles, it was the town of Mansfield that caught the silk fever more than any other.

By 1767, William Hanks, whose family was prominent in the silk business for nearly 150 years, raised enough silk to make three dresses, in direct violation of the British ban on manufacturing in the colonies. He was also able to provide 3,000 mulberry trees for sale. In 1788, a group of some thirty Mansfield citizens petitioned the legislature for a charter, stating "they have been able to raise large quantities of raw silk some of which has been manufactured into cloth and the remaining part into sewing silk which if properly made is equal to anything imported." Hardly a typical business corporation, the members of the Mansfield company lived close together and apparently desired incorporation for the purpose of making bylaws "for the well ordering and regulating themselves, in and about the raising and manufacturing of silk." It proposed to meet annually and to choose a director, treasurer and two silk inspectors. The lawmakers exempted it from any assessment on profits for a period of twelve years. Despite the production of 200 pounds of raw silk yearly, worth five dollars a pound, the company languished. One of the petitioners, Constant Southworth, told Alexander Hamilton's agent in 1791 that "no special advantage can be derived from this grant, however generous, until workmen can be obtained skilled at least in some one branch of the Silk manufacture." One woman and two or three children could make about two pounds of raw silk in a week. It was sold mainly for stockings, ribbons, handkerchiefs, buttons, fringes and sewing silk.

An elderly Mansfield lady has left a description of the tedious process of raising worms and reeling silk as she saw it around 1830:

> The worms hatch out of very small eggs that the millers lay on paper; they are then put in a very cold place and kept until spring. When the mulberry trees commence to leave out, the silkworm eggs are brought out in the case they have been kept in and placed in the living-room. In a few days the eggs will begin to turn dark; they are then going to hatch. Soon the worm is about one-quarter inch long. They are fed upon very tender leaves, and as soon as they crawl on the leaves they are moved onto a clean paper and . . . moved every day until they are all hatched out. They shed their skin four times; after . . . the fourth time, they grow very large very fast until they are nearly four inches long. They are then put on shelves made for them, and small bushes put up for them to wind their cocoons on. After they are done winding, the balls are picked off the bushes; there is loose silk on them which is called tow; that must be all picked off. Then they are put in hot water. Then take a brush and take up the end of the silk and reel it off. The reel is two yards around. When the silk is dry it is stiff and gummy so it has to be put in hot water and soaked some time; it is then spun and reeled into skeins; 20 threads around the reel make a skein of silk . . . Girls picked most of the leaves, it was very hard work and very small pay; they had 10 cents a bushel for picking. Some could pick three bushels a day; the people exchanged the silk for goods at the stores. . . .

In 1810, Rodney Hanks and his son Horatio had a fling at making silk thread and twist by waterpower. In a tiny white mill, no larger than twelve by fifteen feet, they installed a double wheelhead. Previously, silk had been reeled from the cocoon on the large wool wheel, requiring many turns to achieve the necessary twist. The Hanks device, made in a nearby blacksmithy, had gears which speeded up the process. But their venture was short-lived. The Hanks family also built a larger mill at Gurleyville in 1814 and another at Mansfield in 1821, but none proved successful until the formation of the Mansfield Silk Company (incorporated, 1829).

From Mansfield the silk industry spread to Willimantic, which won renown for its thread, and eventually to South Manchester, where the Cheneys created a family empire. By 1831, Windham and Tolland counties turned out 75,000 yards annually on fifty looms. In the Hartford area, Christopher Colt, Jr., a brother of the illustrious Samuel Colt, ran the Connecticut Silk Manufacturing Company, a chartered corporation, having 100 looms worked by girls. During the summer, for a wage of fifty cents daily, they wove silk goods and in wintertime made hats and bonnets out of Tuscan straw. Farmers were encouraged to raise a few bushels of cocoons, have their daughters reel the silk, and furnish it in its raw state to the company, which for five dollars furnished the simple reel itself. This arrangement was a good example of the intermediate stage that frequently existed between some homespun industry and integrated manufacturing.

By 1800 Connecticut had granted forty-five corporate charters to various ventures. Mansfield's was the only one devoted to manufacturing. In the entire United States there were only eight charters for manufacturing. Of Connecticut's other charters two were held by insurance firms, three by toll bridges, five by banks, six by public services like water supply and twenty-three by turnpike companies.

Connecticut led in the number of chartered turnpikes. Crudely constructed, they were still kept in better repair than town highways. The best one, costing $2,280 a mile, was the two-lane stretch from New Haven to Hartford. Its annual income was $3,000. Tolls were collected at ten-mile intervals. Turnpike companies had the right of eminent domain for their routes, but profits were restricted to a maximum of eight percent. By 1850, sixty-nine were still in operation.

New England granted more charters than the rest of the country, Massachusetts being first with sixty and Connecticut next. Charters, then, were rather sparingly given for businesses that would directly contribute to the public welfare. There was no body of corporate law as such. In fact, because they conferred special privileges, the seeking and giving of charters met with resistance right from the beginning. So innocent a would-be corporation as the Connecticut Medical Society in May, 1787, was denounced in the legislature as "a combination of the doctors . . . directly against liberty . . . a very dangerous thing . . . a monopoly." A few years later the magazine *Anti-Monopolist* attacked the Society for Establishing Useful Manufactures:

> This propensity for corporations is very dangerous to the liberties of a people . . .
> we shall inevitably have all our wealthy citizens whether they be engaged as
> merchants, manufacturers or speculators, formed into corporate bodies, to
> aggrandize themselves, and increase the influence of government.

The article described the Society as an undemocratic combination of ambitious politicians and avaricious merchants; as it turned out, the Society failed because of inadequate capital, speculation and both poor machinery and management.

What kind of men founded successful manufacturing concerns in these frenetic years? According to one historian who studied the formation of 140 such companies, "the founders evolved from middle class backgrounds but were endowed with the advantages of location, education, training and opportunity, with a large measure of luck . . ." They started out as farmers, merchants, craftsmen or small entrepreneurs. Merchants commonly went into textiles. Most began work before they were nineteen years old, half of them founded companies while still in their twenties, and more than one-third received financial help from relatives. "In the early period all the founders were either wealthy merchant-capitalists or those who had been apprenticed as youths and became master workmen . . . The basic advantage was not wealth, as most were middle class, but in their upbringing in a practical business environment, high educational attainments and close association in the area of ultimate manufacturing success." Generally, they organized partnerships, scrounging the hard-to-find capital from their families, neighbors, or other businessmen.

But soon the fledgling entrepreneurs sought the advantages of incorporation to meet their needs for more capital, to spread the risks of manufacturing, to limit the liabilities of ownership, and to insure perpetuity of the companies. Most began with ten or fewer stockholders and less than $50,000 of capitalization. The percentage of failure was high. Except in a few isolated cases, profits were insufficient to provide the working capital needed for growth and survival. New funds for carrying on a business had to come from partners, from loans advanced by wealthy merchants or reluctant banks, or from forming chartered joint stock companies. Only one manufacturer in Connecticut, Eli Whitney, owed his existence to government contracts, and for him incorporation never became necessary. Lotteries ceased to be a source of capital after their prohibition in 1828. Thus, after 1810 manufacturers, as well as other businesses—especially banks— flooded the legislature with petitions for charters. Typical examples of early incorporations were David Humphreys's woolen factory in 1810; Christopher Colt's "Hartford manufacturing company" in 1814; Normand Knox's cotton and woolen mill on Roaring Brook in Glastonbury, also in 1814; Charles Starr's "Hartford stereotype company", 1815; and Henry Seymour's "Hartford city manufacturing company" on Mill River, cotton and woolen goods, 1819.

Despite the encouragement given earlier to silk growers like William Hanks and to wool manufacturers like Elisha Colt and David Humphreys, by and large the politicians did not feel manufacturing needed any special assistance. Governor John Cotton Smith assured the legislature at its spring session in 1813:

> Amidst the serious embarrassments occasioned by the war and the antecedent restrictions upon commerce, we have the consolation to witness a remarkable progress in manufactures, and in the cultivation of the useful arts.

Even the promotion of agriculture, especially the raising of merino sheep, hemp and flax, was summarily rejected.

Governor Smith, it turned out, painted much too rosy a picture of economic conditions. Peace in 1815 brought grief to manufacturing, albeit temporary. England, more industrialized than ever, resumed competition by dumping goods on our shores at prices below domestic costs and otherwise doing everything possible to destroy American enterprise. For three years depression gripped the Northeast. Cotton and woolen mills all but collapsed. Factory villages stagnated. To save themselves, mill owners turned to Congress for relief. The result was the passage of the moderate tariff of 1816, which levied a duty of twenty-five percent on cotton and woolen imports. Domestic production revived and gradually squeezed out all but the cheapest fabrics from abroad. The number of Connecticut textile mills more than quadrupled, rising from 29 to 133 in 1818.

In 1817, a group of businessmen, younger merchants and manufacturers, got together in Middletown to see how the still embryonic industries of the state could best be preserved and to seek further legislative relief from foreign competition. Prominent among them were S. Titus Hosmer, Commodore Thomas MacDonough, Elijah and Nehemiah Hubbard, Simeon North, John Alsop and Henry Lyman. In February, they formed the Connecticut Society for the Encouragement of American Manufactures, which many years later became the Manufacturers Association of Connecticut. Denying that noisy factory jobs and crowded boarding houses propagated immorality or swelled the number of paupers, as some Federalists feared, they contended that the promotion and protection of industry were indispensable to the prosperity of New England. Its generally poor soil, migrating population and the decline of commerce left no alternative.

"Few countries," these gentlemen pointed out, "are so well watered as New England—it is full of streams excellently calculated for mills and machinery of all kinds. New England must either manufacture or dwindle into insignificance . . ." They also called attention to the fact that manufacturing had given England a vast commerce, supremacy of the seas and colonies everywhere, and they foresaw a lucrative market for manufactured goods in the Southern states, where cotton was becoming king. The preamble to the Society's constitution stated:

> If New England regards her true policy, she will cultivate American manufactures. They will retain and increase her population, and her wealth . . . We have no necessary dependence upon Europe, nor the West Indies . . . (Manufactures) give employment to the poor; and the means of industry to the idle; furnish a market for the farmer's produce, and become an *aid* instead of a *rival* to our commerce . . .
>
> The state of Connecticut enjoys preeminent advantages for manufacturing establishments. It has a comparatively numerous population—ingenious artisans—industrious habits—sufficient capital—excellent millseats . . . They are the last, best hope of Connecticut.

The Society proceeded to operate on a very informal basis, conducting a survey of industrial conditions by county, memorializing Congress for adequate protection and keeping an alert eye on state legislative trends.

Soon after 1820 industry in the Northeast had solidified sufficiently to mount a lobby in Congress. It succeeded, in 1824, in getting the first real protective tariff. Cotton spinning could now stand on its own feet. Significantly, Northern farmers joined with manufacturers to beat down the free-trade Southern cotton planters. Riding high, the protectionists rammed through exorbitant duties, including fifty percent on woolen goods—the highest until 1860. North Carolina cried out that the tariff of 1828 was "artfully designed for the advancement of the incorporated companies of New England." Even a New Hampshire senator bemoaned the death of homespun industry. Cotton factories were doing well, but the woolen mills lagged behind. Beginning with Henry Clay's Compromise Act of 1833, the tariff forces had to retreat. Except on cotton, there is little evidence, however, that the ups and downs of protection acted as a major stimulant of industrial growth during this period.

To many Congregationalists the moral aspects of manufacturing took precedence over the arguments for economic independence. They saw manufacturing as an evil that made the few rich and the poor poorer. Mill owners inadvertently intensified these anxieties by insisting on reprinting grim tales of English factory conditions as part of their campaign for a protective tariff. Generally speaking, however, Connecticut manufacturers seemed to accept a moral obligation for the welfare of their workers, albeit in a paternalistic and even autocratic manner, but they resisted efforts of the anti-factory forces to legislate corporate responsibility for guarding worker health, insuring good conduct or providing churches, libraries and schools.

Labor, too, was beginning to discover the advantages of organization. Mechanics' associations were the earliest form. Dominated by masters or owners, they also included journeymen and apprentices. Their preoccupation was with the education and welfare of their members. "New Haven probably had such an association in 1793, when a mechanics' library was established," according to Nelson R. Burr. Later, in the same city the General Society of Mechanics was incorporated (1807). Until its liquidation in 1841, it gave loans to apprentices and donations to paupers. Similar societies appeared in Norwich and Hartford. They had, however, little or no impact on the more class-conscious movement for reform during the 1830s. Starting in Philadelphia and New York, this brief spate of militancy spread to New England, attracting the support of farmers and upper-class sympathizers as well as workingmen. It sought the abolition of imprisonment for debt, mechanics' lien laws, the ten-hour day, improved factory conditions, universal suffrage for men, reform of the militia system, free schools for children of the poor and the end to corporate monopolies.

In Connecticut the antiestablishment political activity centered around New London, where a prolabor newspaper, the *Connecticut Centinel*, was published. New London County workers successfully elected candidates for the General Assembly in 1830. The Hartford *Times* encouraged "Workies" to amalgamate with Jacksonian Democrats and declared in favor of the ten-hour day. Local trade unions, suddenly more interested in wages and working conditions than benevolences, were organized. One was the Journeymen Carpenters and Joiners Society of Hartford (1836). Dr. Charles Douglas, a New London physician, was a leading light in New England's labor movement for twenty years. He edited the *New England Artisan* and helped to form the first national organization of wage earners, the National Trades Union, which survived only until the panic of '37 ended union activity.

To a limited extent the "Workies" were successful in pushing reforms through the Legislature, aided and abetted by the Jacksonian element. The Hartford *Times* also lent its support, particularly attacking the issuance of small notes by banks. It pointed out that farmers and laborers were the chief sufferers from bank failures and the depreciation of paper money. "It flayed the corporations as destroyers of democracy and breeders of aristocracy . . . Before 1840, the lawmakers capitulated to nearly every 'Workey' demand, providing for greater popular control of the Legislature, popular election of various public officials, and a more democratic selection of juries . . . The greatest triumphs were scored in 1836 and 1837, in the passage of a limited mechanics' lien law and the abolition, under certain restrictions, of imprisonment for debt."

A constitutional amendment of 1845 eliminated the property qualification for voting. In 1842, the use of child labor was curtailed. Employers were forbidden to work children under fourteen more than ten hours a day in cotton or woolen mills and had to see that they received schooling for three months every year. This was strengthened in 1855 to outlaw child employment under nine years of age. At the same time the ten-hour day became state policy, unless companies and employees agreed otherwise. All in all, however, the record of reform amounted to a meagre return for the effort expended.

* * *

The year 1817 saw the election by a narrow margin of Oliver Wolcott Jr. as governor. This distinguished citizen was a masterful mediator, highly cultured, with a vigorous and imaginative mind. He had turned against the backward-looking "Standing Order" and run as a Republican or Tolerationist. The adoption of

Governor Oliver Wolcott, Jr.

the new Constitution of 1818, which he engineered, marked the end of the Federalist party and the religious monopoly of Congregationalism. It amounted to a social revolution, bloodless to be sure, yet destructive of the old ways. Not only were church and state separated, but religious freedom was guaranteed, suffrage was liberalized, and the door opened wide to industrialism.

For a decade Wolcott presided over a penny-pinching, do-little legislature. Most of his social and economic views were too progressive to be palatable to an assembly composed largely of rural-minded conservatives, regardless of party. Worried over "the numerous emigrations of our industrious and enterprising young men" because of the stagnation of commerce, he urged the lawmakers to exempt textile employers from property taxes and their workers from paying a poll tax or serving in the militia. Once and for all, he wanted settled the question of whether manufacturing establishments should be encouraged by the state.

In this instance they cooperated. Wearing domestic woolens as a patriotic gesture, they also passed a resolution beseeching the people to purchase and use locally-made cotton and woolen fabrics in preference to those of foreign countries. In June, President James Monroe further boosted the cause of industry by stopping off at Whitneyville to see Eli Whitney's armory and, accompanied by David Humphreys, inspecting the mills and shops in Middletown.

One year later Governor Wolcott asked for efficient regulations for "rendering the managers and agents of corporations responsible for the faithful execution of their trusts." Asserting he had no personal knowledge of any abuses, admitting incorporations had been beneficial, yet he observed that the practice of granting charters had "produced great changes in the relations of property, and the ancient modes of transacting business in our country . . ." In addition, he recommended an updating of the law of partnership:

> As this law exists at present, no person can advance money to be employed in business . . . without thereby subjecting his whole property to responsibility. The effect is, that men in advanced age, or retired from business, are restrained from employing their capital in . . . undertakings which would be highly beneficial to the community . . .

> Consider whether it is not contrary to public policy, if not an abridgement of private right, to restrain individuals from forming partnerships with a limited responsibility. As the community are interested in being guarded against frauds arising from secret combinations, it would be proper to require contracts of this nature to be recorded in a public office . . .

Wolcott's second suggestion was warmly debated in the May session. A Mr. Stevens said he thought the bill a very proper one; he did not know why other incorporated companies should be more restricted than banks; stockholders in banks were not liable in their personal or private property for the debts of the company. If a change were not made, capital would be withheld, and young men of enterprise could not do business to advantage. And so an important step toward modern corporate life was taken with the passage of the bill to provide limited liability to stockholders who risked their savings to create new enterprises. Except as it initiated a body of corporate law, the legislature nonetheless continued to adhere closely to the doctrine of laissez faire.

In the ensuing years Governor Wolcott expressed concern over the nation's defective system of finance and the lack of protection for its internal industry:

> Our industry is becoming languid; our currency consists of notes . . . due to banks; even usury is less profitable than heretofore; our national debt exceeds what it did when the present government was first organized . . . Our agriculture, commerce and manufactures are equally depressed . . .

The answer, he reiterated, lay in promoting the productive powers of industry. Agricultural societies, sheep raising, better farming methods, in his opinion, would not assure prosperity; instead, priority should be given to the introduction of machines to turn hemp and flax into the finest fabrics. In 1822, he also pushed for a mechanics lien law to help workers recover unpaid wages, but was again rebuffed.

The currency crisis still prevailed when the Whig Governor William W. Ellsworth took office in 1838 and temporarily halted the growth of manufacturing. "Who is to answer," the Governor asked, "for the prostration of credit . . . the breaking up of families, the loss of property and the despair seen around us in our commercial cities and manufacturing villages? . . . The manufacturers of New England . . . are baffled, crippled, and desponding beyond endurance; they cannot sell what they have manufactured, nor obtain remittances for that which they have sold: hence their establishments are closed, their hands dismissed, their plans deranged . . . and too many of the young and enterprising, deserting their native soil, are seeking occupation in the West." Connecticut was enmeshed in the fear and gloom caused by the financial panic of 1837.

Amid the depression there passed a monumental piece of legislation which more than offset the dismal record of accomplishment by the standpat lawmakers, the majority of whom had otherwise closed their minds to the reform movement of the Jacksonian era that had swept the rest of the country. This was the 1837 "Act relating to Joint Stock Corporations", often called the Hinsdale Act after its chief draftsman. It not only marked the real beginning of the modern business corporation in Connecticut but provided the foundation for similar laws in most other states and was imitated in England as well, when Parliament passed the

Limited Liability Act of 1855. It preserved the essential features of the code of corporate behavior which Lord Coke, in 1612, drew up as a guide for Queen Elizabeth in the franchising of English mercantile ventures. More than any other piece of legislation ever enacted by a state body, it had, as Odell Shepard noted, "a profound effect upon industrial development throughout the world."

At the time it also struck a powerful blow in behalf of equal rights and free competition by ending the age-old practice of granting charters. Stimulated by the greater freedoms resulting from the new Constitution and the waning influence of the "Standing Order," the rising class of business adventurers together with democratic politicians cried out against the charter system. Since influence was essential to obtaining a corporate charter, and since once granted the favor conferred a virtual monopoly, abuses were bound to occur. Apprehension was expressed over the proliferation of banking charters in particular, because they carried to excess the exclusive privilege of issuing bills of credit. Ugly rumors circulated of kickbacks and bribes to key legislators for support of such special acts. It was patent that those promoters who were the most popular and frequently the most generous with the legislature succeeded in getting charters for their pet ventures, while the nobody got nowhere. Charters were often vehicles for speculation and exorbitant gain, with the public being the goat in case of adversity. Thus, the chartered corporation became in the public eye synonymous with power and wealth and the foe of the freeman and small manufacturer.

The battle for a general corporate law in Connecticut pitted larger towns against small ones (which dominated the legislature)—mill hands against artisans, liberals against the vested interests and, at first, oddly enough, against a Democratic governor, who in all other respects behaved more liberally than his predecessors. An able lawyer and politician, the grandson of that firebrand preacher Jonathan Edwards, sometimes called the "Pope of the Connecticut Valley," Henry W. Edwards had been elected governor first in 1833 and again in 1835, remaining in office until 1838. The Democrats controlled the General Assembly for the first time. The conservatives wanted no reforms, supported high tariffs and espoused the cause of manufacturing. The chief protagonist for reform was the balding, bearded young representative and Democrat from Glastonbury, Gideon Welles, who occupied two influential positions. From 1826 to 1837, he edited the Hartford *Times*, the official organ of Jacksonian democracy in southern New England, and in the General Assembly sat on the "Joint Standing Committee on Incorporations other than Banks." Later as an abolitionist he turned Republican and became President Lincoln's loyal secretary of the Navy and confidant throughout the Civil War.

Gideon Welles

At the 1835 session Welles had presented a report from his committee which cognetly stated the case for a general law of incorporation yet stopped short of complete endorsement:

> It has become a grave and serious question . . . whether grants of private incorporation are conducive to the public welfare, or compatible with the spirit of liberty . . . Our legislators for years have been building up artificial distinctions by special and privileged laws . . . The operations of this system are oppressive on individual enterprise . . .
>
> A corporation is correctly defined to be "a body politic, or artificial person, of capacity to grant and receive, to sue and be sued, maintaining a perpetual succession, and enjoying a kind of legal immortality." Although the legislature of this State has, for some years past, adopted the policy of establishing these soulless and everlasting bodies, to an almost unlimited extent, your committee look upon it as inconsistent with the spirit of our institutions, and of a dangerous tendency . . .
>
> A large portion of the time of each session of the General Assembly is now consumed in granting these favors to those who are no more entitled to them than their fellow citizens. If the powers and privileges thus conferred are necessary for our welfare, or, if essential to our prosperity, it is suggested whether a general law . . . would not be as well for the petitioners . . . Your committee can see in it many advantages over the present, pernicious, imperfect and partial system.
>
> A general law . . . would abolish the monopoly, by opening the door to all who might wish to avail themselves of it.
>
> It would destroy the present demoralizing system of granting special favors . . .
>
> It would put an end to perpetuities, the worst feature in our corporation system . . .
>
> The unobtrusive work-shop of the mechanic, the residence of freedom . . . cannot compete with incorporated wealth . . . The avenues to wealth are blocked up by your corporation laws . . .

As a result, the General Assembly rejected the numerous petitions for charters that had been filed that year, stemming the further spread of state-approved monopolies. But Welles had planted the seed of change and now prepared for the next round of battle the following spring. Just before the 1836 legislature convened, he wrote in the *Times*:

> The majority are convinced that the granting of individual privileges is at war with the doctrine of "equal rights"—that a free competition . . . is the only principle compatible with free institutions.
>
> Charters should be obtained under a general law, which would destroy the monopoly, and allow the mechanic who enters upon business with no resources

but his trade, to enjoy an equal privilege with his neighbors for acquiring wealth
. . . Let Connecticut . . . take a stand in advance of her sister States, in this work
of radical reform. . . .

In contrast, by questioning the good Jacksonian doctrine of corporate privileges
for every one, Governor Edwards unintentionally put himself in the enemy camp:

> The affairs of corporate companies are not generally as well managed as those of
> individuals. Their business must be done by agents who have not the same interest
> . . . The exemption from loss beyond a certain sum, has a tendency to create
> boldness in undertakings, and improvidence in the mode of conducting them . . .
> In those branches of business which can be and are carried on by individuals, no
> good but much evil will probably result from acts of incorporation. There are some
> works which require more capital than individuals would furnish. Of this nature
> are canals and railroads.

On May 17, the nine-man Committee on Incorporations reported out a bill
embodying Welles's principles of free and equal incorporation "for all joint stock
manufacturing companies, voluntarily associating in numbers not less than five;
stock to be not less that $10,000, nor more than $200,000; all shares fifty dollars
each, a certificate of amount of stock to be deposited in the office of the Secretary
of State." It was read a second time on May 23; tabled; amended the next day and
tabled again; read again on May 26 and amended to cover mercantile as well as
manufacturing firms and to reduce the minimum number of stockholders to
three, the capital required to $2,000 and the price per share to $25. There was
considerable objection to including commerce, a Mr. Sterling contending that it
lacked "the grand characteristic of manufacturing business, which was that of
producing the material of trade; while the commercial and mercantile business was
merely speculation . . ." An attempt to make each stockholder liable for the
debts of the company failed. The debate continued the following day. Mr.
Rockwell, who had introduced the original bill, was now fighting hard for its
passage. Joint-stock corporations, he argued, were institutions wholly confined to
republican countries. In England and Europe they were almost unknown. Here
they had been liberally extended, bringing—among other benefits—lower
insurance rates and industrial growth. There was momentary silence. The yeas
and nays were counted, and the bill lost by a vote of 104 to 94.

The bill fared no better in the Senate, despite the eloquent plea of Senator
Martin Strong. He saw the new law as a way of eliminating the burden of
petitions imposed on the legislature over the previous decade by those seeking
charters. He believed manufacturers received too little encouragement. "Pass this
law, sir, and you will see a healthy influence extended over our State, our
numerous waterfalls will be covered with manufacturing establishments, villages

will spring up, and our enterprising inhabitants will find employment . . ."

His was a prophecy that came to pass, but the law itself had to wait until the 1837 session. Governor Edwards and the farmers had won the second round; Mr. Rockwell, Gideon Welles and other members of the Committee on Incorporations were dejected. The *Courant* commented: "The poor debtors . . . have been left to get out of prison as they can . . . Corporation privileges are not yet extended to the whole community by a general law, but remain as formerly, dependent on special legislation . . ."

The following spring, Edwards defeated the Whig candidate, William W. Ellsworth, son of the distinguished jurist, Oliver Ellsworth of Windsor. His reelection apparently resulted in a change of heart regarding joint stock companies. Somewhat cryptically he told the legislators that "our country, from the agricultural, is rapidly advancing to the manufacturing state" and warned that unless the laws were changed, industry "will be compelled to look elsewhere for an asylum." He made no mention of vote buying under the charter system but accepted the argument that reform would eliminate an excess amount of private acts and save legislative time.

In this critical year of widespread depression, the efforts of Gideon Welles were reinforced by the election of Theodore Hinsdale of Colebrook, who served only one term as a representative but made the most of it. Hinsdale was related to Daniel Hinsdale, the Hartford merchant who a half century before had been a stockholder in the Hartford woolen manufactory and also its agent. A graduate of Yale in 1821, he read law for a while, then with his father-in-law took over an established scythe business under the name of Rockwell & Hinsdale. His father, a Winsted merchant who bought and slaughtered cattle for the West Indian trade, had two brothers who started a small unincorporated business in Middletown; when they suffered financial reverses, he agreed to endorse their notes to enable them to obtain a loan. Bankruptcy was the unhappy outcome. Hinsdale not only saw his uncles go under, but his father's business and personal property swept away.

Out of this family tragedy Hinsdale became convinced of the small businessman's need for a legal device, easily procured, properly safeguarded, to encourage investors to put their savings in promising enterprises without having to face ruin in the event of failure. The chartered corporation he saw as too cumbersome, too subject to favoritism, too regardless of the public's interest, too special a mechanism to accelerate the growth of trade and industry in a responsible manner.

Although Hinsdale did not serve on the corporation committee with Welles, he worked behind the scenes to polish the provisions of the model law, borrowing liberally from the Massachusetts regulations of chartered corporations, using his commanding personality, abundant energy and oratorical power to overcome his rural opponents, who generally regarded industry as an evil upstart and corporations as commercial kingdoms. His achievement in drafting so perfect a piece of pioneer legislation and in helping Welles steer it through committee and both branches of the General Assembly were the crowning points of his career. Four years after his victory he succumbed to typhoid fever at the age of forty.

The final draft of the legislation was reported out of the house on May 26, tabled long enough to print 500 copies, and on June 2 called for another reading on a motion by Welles. Action was delayed, however, until June 6, when it passed 108 to 84. Despite the size of the opposition, there was little debate or dissension and a remarkable omission of comment in the press. The significance of its enactment was lost in the public hue and cry over the state's financial woes. The suspension of specie payment, permitting the banks to circulate small denomination paper notes, the abolition of imprisonment for debt at last, as well as Edwards's controversial veto of the repeal of a bank charter in New Haven, were of more immediate concern. By the spring of 1839, fortunately, the economy turned around, and the new governor could report, with political hyperbole, that "the dark cloud is dispelled; confidence has revived; individuals invest their capital without fear . . . and a sound and convenient currency imparts new life to business, gives encouragement to the manufacturing village and casts the sunshine of cheerfulness over the commercial city."

Connecticut's corporation act, the substance of which was Hinsdale's but to which Welles brought life, provided for most of the statutory powers peculiar to what has become the sine qua non of business today: to sue and be sued, to have a common seal, to elect officers, to fix their compensation and duties, to establish bylaws, to employ agents, mechanics and laborers. Incorporation was limited to one purpose, to be distinctly set forth in the articles of agreement which all incorporators had to sign. A board of directors and the officers of president, treasurer and secretary were specified. The corporation could forfeit stock not paid for; it had a lien upon the stock of its members for debts. Upon organization the officers were required to file with the secretary of state and the clerk of the town where the company intended to operate a certificate setting forth its purpose, the amount of capital stock, the names of stockholders and the number of shares held by each. Annual reports were obligatory. Stockholders were made liable for all capital refunded to them and for the declaration of illegal dividends. Altogether, it contained twenty-three sections.

Two final but futile attempts were made by the conservative Whigs to restore the charter system. Newly elected Governor Ellsworth, although recognizing the vital role of manufactures in making New England prosperous and in stemming emigration, still could not understand the public odium against chartered companies:

> Have (they) not paid labor with promptness, and with less default than private associations of equal business? Have they particularly practiced frauds on the community, or trampled on the rights of the citizen? The reckless attacks . . . have filled capitalists with apprehension, and shaken the foundation of the social fabric . . . An indiscriminate outcry against them as monopolies, as exclusive privileges, as aristocratic, as combinations of the rich to impoverish the poor, is mischievous . . .

Several years later Governor Roger S. Baldwin argued that this "experiment in legislation" had resulted in fraud and speculation. He feared extending chartered powers to every kind of business while requiring no personal responsibility on the part of stockholders would soon prove the law evil. In one thing he was right: the "Standing Order," which had so long enjoyed both political power and economic privilege, had suffered another body blow. The doors of competition had been opened wide. But few lawmakers wanted to fight to turn back the clock, and the governor's doubts fell on deaf ears. Under the new law 44 joint stock companies were formed from 1837 to 1840; in the next decade, 142 were incorporated; and during the 1850s the number soared to 548. By the time of the Civil War the corporate era had definitely arrived.

CHAPTER VI

Clocks for Every One

AMERICA in the nineteenth century swelled westward from her seaboard. Axe and gun cleared the way for restless, land-hungry settlers, while Connecticut clocks made their rigorous life orderly and efficient. Clockmaking, by means of the acquaintanceship between the two great "Elis" of invention—Whitney and Terry—became the second industry to adopt the concept of interchangeable parts.[1] The immediate result was to slash the price of timepieces so that every cash-poor family could purchase what had always been a status symbol of the well-to-do.

For clockmaking to adopt the technical revolution in gunmaking meant the entrenchment of mass production and its application in rapid succession to sewing machines, machine tools, bicycles, typewriters, automobiles—everything, in fact, that constitutes our material well-being today, products not only obtainable in quantity at low cost, but also generally of better quality than could be made by hand.

The taken-for-granted common clock was perhaps the first luxury to become a necessity. By the 1840s a travelling English scientist wrote home that "here in every dell in Arkansas and in every cabin where there is not a chair to sit in, there is sure to be a Connecticut clock." Yet forty years earlier, even in the clockmaking town of Bristol, only a handful of residents could afford to own a clock.

Until Eli Terry, Seth Thomas, Joseph Ives, and Chauncey Jerome arrived on the industrial stage, clockmaking was an art and a craft calling for many skills. Each timepiece, made to order for the discriminating buyer, had more artistic attention given to the case than to the works, usually cast from brass. The

---

[1] The versatile Pitkin Brothers in East Hartford first used factory methods in watchmaking (1837–42). Although unsuccessful, they trained men who later contributed to starting a watch factory in Roxbury, Mass. (1850) that in turn led to the founding of the great Waltham Watch Company.

Eli Terry

eight-day, striking "tall" clock, an expensive decoration, cost as much as eighty dollars. Its selection was as important a decision for those colonial families who could afford one as an automobile would be today. Nearly a hundred self-sufficient clockmakers were scattered about the state during the eighteenth century. Penrose R. Hoopes, a leading authority on the subject, observed "this early and wide distribution of clockmaking in the country towns of Connecticut, and the resulting self-reliance and enforced readiness of the clockmakers to turn a hand to any variety of mechanical work . . . fostered the ingenuity and commercial aggressiveness which ultimately resulted in the preeminence of Connecticut as the center of clockmaking in America."

Thirty-hour clocks with wooden works of oak, cherry or laurel were almost wholly a Connecticut development. The earliest, 1745–50, were probably the inspiration of Isaac Doolittle of New Haven or Benjamin Cheney. The prolific Cheneys made East Hartford a center for this business. For a short while John Fitch, later the inventor of the steamboat, worked for Benjamin. Eli Terry of Plymouth learned from either him or his brother Timothy techniques that he subsequently incorporated into his design of the wooden shelf clock. After the Revolution, men like Gideon Roberts of Bristol, James Harrison of Waterbury and Terry followed the trail of tinware peddlers selling wooden clocks with pendulums throughout the New England and Middle Atlantic states. Roberts and his four sons sold several hundred tall clocks a year and maintained a warehouse in Richmond as well. He subcontracted the cases. Chauncey Jerome, in his autobiography, said: "He was an excellent mechanic and made a good article. He would finish three or four at a time and then take them to New York State to sell. I have seen him many times, when I was a small boy, pass my father's house on horseback with a clock on each side of his saddle and a third lashed on behind . . ." Just before his death in February, 1813, Roberts wrote his sons in the South: "I have parts for 80 clocks which I mean to put together by May 1st, and I have been collecting lumber for 1000 more this winter . . ." These quantities and his selling price of twenty-five dollars indicate that Roberts thought in mass production terms.

Terry also loaded his saddle twice a year with clock movements and trotted off into the hinterland west of the Hudson River. His price, the same as Roberts's, included a brass dial, second hand and the moon's phases. Often it was paid by the farmer in cloth, beeswax, hogs or salt pork. Sometimes the buyer was trusted to pay in installments. If he was too lazy to furnish his own case, he simply hung the mechanism on the wall, whence came the name "wag-on-the-wall."

Two other craftsmen of this period had a lasting effect on the clock industry. In 1773 Thomas Harland of London boarded the ship from which the famous tea was dumped into Boston harbor and opened a clock and watch shop in Norwich near the store of Christopher Leffingwell. Harland is credited with being America's first watchmaker. For the times his was a large operation employing numerous apprentices. One protegé was Daniel Burnap, who started his own place in East Windsor about 1780 and soon advertised "Brass Wheel'd Clocks" for sale. Burnap also excelled in turning out surveyor's instruments, shoe buckles, silver spoons, jewelry and saddlery hardware. Six years later, just after his fourteenth birthday, Eli Terry, born in the same town, bound himself to Burnap. It is possible that before finishing his apprenticeship he spent some time with the master Harland. By 1815, when Terry was making shelf clocks by the hundreds, mostly by machine, the pioneers of the grand cast-brass clocks had passed from the scene.

The true measure of Eli Terry's place in history has been given us by another of Connecticut's great clockmakers. Chauncey Jerome wrote of Terry:

> Eli Terry was a great man, a natural philosopher, and almost an Eli Whitney in mechanical ingenuity . . . He was the great originator of wood clockmaking by machinery in Connecticut.

Modest and temperate, neither smoker nor drinker, always the keen-eyed craftsman, Terry looked the very model of a Yankee. His was certainly an Horatio Alger career. He left Burnap and East Windsor in 1793 for the tiny village of Plymouth, most likely because his girl friend, Eunice Warner, lived there. Two years later he married her, and she subsequently bore him nine children. For a while he made tall clocks in dozen or so lots out of both brass and wood, "having a hand engine for cutting the teeth or cogs of the wheels and pinions, and using a foot lathe for doing the turning," as well as cutting some parts with a knife. In 1797, he conceived his equation clock to show the difference between mean and apparent time. The patent was signed by President John Adams, but the public did not give a tinker's dam whether time was based upon the sun, moon or perfection. Apparently, the only one he ever sold was to the Center Church in New Haven, causing a local furore over what was denounced as "a public nuisance."

Around 1802, while Whitney was perfecting his system of interchangeable parts in Whitneyville, Terry decided to abandon peddling and concentrate on manufacturing. Everyone, he reasoned, needed a clock but only one in ten could afford the price. The obvious answer was to find a way to cut the cost to fit the slimmest wallet. Terry built a small plant on a stream so that he could use more machinery and turn out wooden clocks, not by the dozen, but by the thousand. His neighbors thought he had lost his senses and ridiculed the idea of his being able to find a market for so many.

Heedless of criticism, the inventor in 1807 negotiated an unprecedented contract that plunged him irretrievably into mass production and in a sense united the clock industry with the emerging brass industry. To the amazement of all his business associates, Terry agreed to provide the Porter brothers of Waterbury with four thousand clock movements at four dollars apiece, probably as many as the entire state had assembled in the previous decade. According to his son Henry, "it took a considerable part of the first year to fit up the machinery, most of the second year to finish the first thousand clocks, and the third to complete the remaining three thousand." One of Terry's innovations was a machine to cut the teeth of his wooden wheels, a process which Hiram Camp, another clock manufacturer, said "was hinted to him by Eli Whitney." Terry's profit was great enough for him to entertain the thought of retirement, even though he hadn't yet reached the age of forty. He sold his factory to his assistants, Seth Thomas of Wolcott and Silas Hoadley of Plymouth, both young carpenters whom he had trained as assemblers.

There followed a confusing succession of changes in partnerships and plant locations which became as common to the clock industry as to its sister brassmaking. Clockmaking also suffered from furious competition, the audacious copying of designs, infringement of patent rights and frequent bankruptcies. The price for movements alone fell to as low as five dollars. Thomas and Hoadley soon parted company, each one making tall wooden clocks under his own name at rock bottom prices. Meanwhile, Terry, consumed by a bold new idea, bought another factory site on the Naugatuck River, and before long there were three Terry firms in Plymouth operated by five different family members. Terry, Sr., spent the years of 1813 and 1814 equipping his second shop with machines to manufacture his masterpiece, a thirty-hour, weight-driven, shelf clock.

According to one employee, this mechanism with its short pendulum, wheels of cherry wood, hand painted dial and glass panel, "completely revolutionized the whole business. The making of the old-fashioned hang-up wood clock . . . passed out of existence." Eli's two oldest sons, Eli Jr. and Henry, now entered the mushrooming business. By 1819 he was completing 6,000 clocks a year at fifteen dollars each, including the twenty-inch high case. Eventually, production reached a peak of 12,000. The patent which he obtained in 1816 was so loosely drawn that it caused him, in his son's words, "no little trouble, strife and litigation." Continued Henry Terry:

> Some of the important improvements which should have been secured by this patent, are in use to this day (1853), and cannot be dispensed with in the making of low-priced clocks, nor indeed with any convenient mantel clock. The mode or method of escapement universally adopted at this time in all common shelf clocks was his plan or invention.

Seth Thomas

Seth Thomas Clock Company, Thomaston

Ironically, the inventor became famous not for the clock's mechanism but for its "pillar-and-scroll-top case," which a former apprentice (Heman Clark) may actually have designed and which others like Chauncey Jerome constructed for him. For a royalty of one thousand dollars he gave Seth Thomas the right to reproduce his shelf clock, and soon Thomas rivalled him in both production and prosperity. About the time Eli Jr. left to build his own factory on the Pequabuck River (1825), his father and Thomas were said to have accumulated a fortune of $100,000 each. Finally, in 1833, at the age of sixty-one, Eli Sr. quit manufacturing, leaving his sons, Eli Jr. in Terryville, Henry and Silas to carry on the family name. Not only did the durable old man survive his oldest son but also his wife of fifty-four years; moreover, he remarried, sired two more children after attaining his three score and ten and lived another decade. With his passing also went his masterpiece, the wooden clock.

The clock business Seth Thomas founded in 1813 in the part of Plymouth now called Thomaston is the only one that has continued until modern times. It is now part of General Time Corporation. Conservative and sharp-tongued, Thomas had no inventive ability, hated to change his methods or designs, and when he did merely copied others. Throughout his life he remained on good terms with Terry, whom he outlived, but detested his other formidable competitor, Chauncey Jerome, because the latter forced him to make bronze looking-glass clocks and to follow Jerome's lead with one-day brass movements.

For the money-mad Thomas, Jerome, in turn, had only contempt and claimed he never made a single manufacturing improvement.

Looking back, it seems rather surprising that Terry, in the light of his association with the early Waterbury brassmakers and his dedication to low-cost production, did not take the lead in converting from wood to rolled brass. That development began with the very clever Joseph Ives of Bristol, who in 1818 invented a metal clock, according to Chauncey Jerome, "making the plates of iron and the wheels of brass. The movement was very large, and required a case about five feet long." Four years later he received a patent on "Looking-Glass Clock Cases," the original of which was signed by Eli Terry as a witness. But not until 1832 did "Uncle Joe" Ives perfect the first practical, eight-day clock movement of rolled brass, embodying the rolling pinion principle to reduce friction. At twenty dollars, it competed with those being made by the Willards.

Ives prospered until the panic of 1837, from which point on he suffered one financial reverse after another. Nevertheless, he made one other major contribution to clockmaking, the so-called "wagon spring" which was used to drive the thirty-hour, eight-day and thirty-day brass clocks. Few of the latter style were ever produced, but they all kept perfect time and were especially popular with jewelers. When his friend Terry patented his shelf clock, Ives hoped he could help make it an even more original invention by substituting a spring-driven attachment for the weights. Had he succeeded, he undoubtedly would have made his fortune. Instead, by the time of his death in 1862, his wagon-spring idea had been replaced by the cheaper coiled steel spring. Unlike Terry, Ives never amounted to much as a businessman, but his contribution alone of the first American rolled-brass clock entitles him to be placed alongside Terry as one of the two most creative inventors in the industry prior to the Civil War.

* * *

In the winter of 1816 a young Plymouth cabinetmaker entered the employ of Eli Terry Sr. to make pillar-and-scroll cases. He soon developed an admiration for the father of clockmaking that lasted the rest of his own eventful life from rags to riches and back to rags. More than any other Connecticut clockmaker of his time, Chauncey Jerome grasped the potential of mass producing timepieces and refined the Whitney-Terry system of manufacturing. His father, a blacksmith and nailmaker, died when Jerome was only eleven, forcing him to leave home and make his own way. In his autobiography Jerome wrote that 1815, the year he married, "was probably the hardest one there had been for the last 100 years, for a young man to begin for himself." Food prices were extremely high, a staple like brown sugar costing thirty-two cents a pound. To Jerome's fascination Terry was

Chauncey Jerome

Jerome's clock factory, New Haven, (formerly Isaac Mix & Son's Carriage Factory).

busy mechanizing the making of pine cases. Soon Jerome also tried his hand at clockmaking in his home. Confident he was ready to make his fortune, he sold his farm in 1821 to his boss for a hundred mantel-clock movements, complete with dials, tablets and weights. Making cases for these and another 114 mechanisms, he exchanged the lot for a house, barn and seventeen acres of land in Bristol, where he moved and set up a small shop. "The first circular saw ever used there was put up by myself, and this was the commencement of making cases by machinery in that town." The idea for the saw, however, he borrowed from Terry. In one long fifteen-hour day Jerome found he could put together four cases. In 1827 he introduced the "Bronze Looking-Glass Clock," with a cheap mirror glass, which could be made for a dollar less and sell for two dollars more than Terry's shelf clock. At first both Terry and Thomas disparaged it to the peddling trade, but the southern market liked it so much his competition was forced to copy him. Jerome had now hit pay dirt.

By the mid-thirties southerners had had their fill of clock peddlers, too many of whom were unscrupulous sharpies. In the case of the Terrys, the label "warranted if well used" indicated the manufacturer's respect for the quality of his product, but not every clockmaker was so finicky. Virginia, South Carolina and Georgia countered by raising the cost of a peddler's license so high as to drive him out. Jerome and his brother Noble were unwilling to lose such a lucrative market without fighting back. To circumvent the license fee, they shipped parts and cases from Bristol to Richmond, where they were assembled and sold as if made in Virginia. Then came the panic of '37. Said Jerome: "Clockmakers and almost everyone else stopped business."

One sleepless night in a Richmond hotel, Chauncey had an inspiration:

> While thinking over my business troubles and disappointments, I could not help feeling very much depressed. I said to myself I will not give up yet. I know more about the clock business than anything else. That minute I was looking at the wood clock on the table and it came into my mind instantly that there could be a cheap one-day brass clock that would take the place of the wood clock . . . I lay awake nearly all night thinking this new thing over. I knew there was a fortune in it.

This may have been a slightly melodramatic version of his brainstorm, because his nephew later stated that Jerome had come across a German brass clock movement in Virginia. Anyway, he and his brother built the original model of what he had in mind early in 1838. He wanted a thirty-hour, weight-driven clock of rolled brass. This was accomplished simply by converting the eight-day movements, such as Ives was making in Bristol. It was an immediate sensation. Besides being cheaper, it overcame the major disadvantages of the wood clock which had annoyed every one for a half century. Wooden clocks could not stand dampness; gear teeth broke easily; and the eight-day models were so noisy as to be called "groaners." Jerome appreciated his achievement better than anybody: "What I originated that night on my bed in Richmond has given work to thousands of men yearly for more than twenty years, built up the largest manufactories in New England and put more than a million dollars into the pockets of the brass makers." Next to Whitney's gun Jerome's brass clock was the first entirely standardized metal product in the world.

Three years after beginning production of his thirty-hour wonder, Jerome, who couldn't stand having partners any more than Colonel Colt, bought out his brother and cleared $35,000. His competitors protestingly followed him out of wood into brass, and he shared with them his know-how, although he claimed that their work was inferior and that, "Yankee-like," they ran the price down. In 1841 he decided to crack the foreign market. Previously, wooden clocks from America could not survive the trip overseas because the works swelled. Even his brass models encountered the usual Bulldog resistance on the grounds that England made the best wooden clocks in the world and brass ones were good for nothing or they wouldn't be offered so cheap. His first consignment to Liverpool was held up by the customs people for undervaluation. English law at the time paid a shipper of confiscated goods the declared value plus ten percent. To Jerome this seemed the easiest way he could imagine to earn cash without any selling expense. After two more confiscations, the authorities caught on, got out of the clock business and let Jerome in.

Still on the ascent, with his name now more widely known than even Terry's, Jerome started a third clock operation in New Haven, buying the bankrupt carriage factory of Isaac Mix & Son and, to the astonishment of the townspeople, running his machines by steampower. His manufacturing ability carried clockmaking to new heights. Parts for cases were processed in lots of ten thousand; dials were automatically stamped; three men in one day cut the wheels for 500 movements; and he reduced his labor cost to a mere twenty cents per clock. Even at the cut-rate price of seventy-five cents he could make a profit. Now he could boast he was the world's largest clockmaker.

Then came his first major setback, or rather three in a row. A profligate clock trader in Litchfield swindled him out of $40,000. Just as he was recovering from a bout of illness, fire destroyed one of his two Bristol factories, in 1845, ruining 70,000 movements. His horde of competitors hopefully whispered he was done for. Instead he enlarged his New Haven plant and cut his costs lower than any of them, making a case alone for as little as twenty cents. Finally, in 1850, he induced the Waterbury brassmakers, Benedict & Burnham, to form a joint stock company. With his financial load lightened, his foreign sales brisk, his

Movement of Terry's original shelf clock

fortune—so he thought—assured for the rest of his life, he turned his business over to his son's care and began to enjoy his New Haven mansion, to travel and to participate in town affairs. Soon, however, he was confronted with the worst crisis of his career.

* * *

One of Connecticut's most acrimonious, tempestuous, disastrous business ventures resulted from the plunge of that master humbug of his day, Phineas T. Barnum, into what was for him the strange sea of finance and manufacturing, drowning forthwith both himself and the Jerome Clock Company. The great showman was born in the back country village of Bethel, near Danbury, in 1810. At age twenty-four, disgusted with the narrow-minded, drab and—in his view—hypocritical attitudes of most Congregationalists, he moved to New York to seek a better living. He loved life and fun and yearned to share these feelings with his fellow men and, of course, at the same time to achieve fame and fortune. He was probably the first American who determined to make entertainment respectable. "This is a trading world," he said, "and men, women and children, who cannot live on gravity alone, need something to satisfy their gayer, lighter moods and hours, and he who ministers to this want is in a business established by the Author of our nature."

It was eight years before Barnum realized his dream of gaining control of the run-down American Museum in New York. Eventually, he made Bridgeport, a fast-growing town of over 5,000, his home. On a large tract of land overlooking Long Island Sound, just as Sam Colt did in Hartford a decade later when his pockets bulged with money, the balding, pudgy, bulbous-nosed teetotaler erected an unbelievable extravaganza of a domicile. Costing $150,000, "Iranistan" must have dazzled the townspeople with its towering center dome, Oriental-style minarets, arches, park, fountain and orchard. Here, in 1855, flush, famous, only forty-five, Barnum retired. Yet before the end of that year his whole world too collapsed.

For some time he had been preoccupied with the promotion of a grandiose scheme to create a new city in East Bridgeport. Having acquired a tract of 224 acres, he "laid out the entire property in regular streets, and lined them with trees, reserving a beautiful grove of six or eight acres, which we . . . converted into a public park . . . We built and leased to a union company of young coach makers a large and elegant coach manufactory." To encourage the purchase of lots, he advanced funds, while holding off the market every other lot on which he hoped to profit from the anticipated rise in land values. Thus, as he admitted, he was a natural sucker for "any plausible proposition that promised prosperity to East Bridgeport."

P. T. Barnum

One of his side interests was a clock factory in Litchfield, which failed. Concurrently, the clock factory belonging to Theodore Terry of Ansonia, Eli Sr.'s nephew, was consumed by fire. Terry induced Barnum to build him a new plant in the East Bridgeport development and become his partner. Then in the summer he received a visit from a Mr. Jerome of New Haven, representing the famous clockmaker. Barnum insisted it was Chauncey himself, but the latter, retired from business and serving as mayor of New Haven, stoutly denied he had ever called on Barnum and claimed it was his son instead. Here is Barnum's side of Jerome's proposition and the ensuing debacle:

> I should lend my name as security for $110,000 in aid of the Jerome Clock Company, and the proferred compensation was the transfer of this great manufacturing concern, with its 700 to 1000 operatives, to my beloved East

Iranistan

> Bridgeport . . . It was just the bait for the fish . . . yet . . . I called for proofs,
> strong and ample, that the great company deserved its reputation . . . .

He found the Jeromes had net assets of over $200,000 but badly needed
fresh capital because of a poor season, a glutted market, $400,000 of clocks on
hand, and notes long overdue. "I was also impressed," he added, "with the
pathetic tale that the company was exceedingly loth (sic) to dismiss any of the
operatives . . . The President, Mr. Chauncey Jerome, had built a church in New
Haven, at a cost of $40,000 . . . he had given a clock to a church in Bridgeport
. . . The Jerome clocks were for sale all over the world, even in China, where
the Celestials were said to take out 'the movements' and use the bases for little
temples for their idols, thus proving that faith was possible without 'works'."
The previous year sales had hit $700,000. Moreover, a New Haven bank,
unbeknownst to him a large creditor, assured Barnum of the firm's financial
strength. Jerome himself guaranteed repayment of the loan.

Thus informed, Barnum signed a series of notes but, incredibly, neglected
on some to fix a date for repayment. Within a month Jerome's son, as secretary
of the company, further lulled him into complacency by writing that it would
soon be in a condition to "snap its fingers at the banks." Barnum put his name to

more notes. Soon he uncovered "the frightful fact that I had endorsed for the clock company to the extent of more than half a million dollars, and most of the notes had been exchanged for old Jerome Company notes due to the banks and other creditors . . . I was a ruined man!" And so he was. Barnum and the Jerome Clock "Bubble" became front-page news everywhere; he lost "Iranistan," his book royalties and his entire fortune except for the American Museum lease—which was fortunately in his wife's name. "In every way it seemed as if I had been cruelly swindled and deliberately defrauded." At the age of forty-six he had tumbled to the bottom of the ladder. One of the few who offered him a hand up was his protégé Tom Thumb, who came from Bridgeport. Barnum rebounded from his misfortune—even managing to capitalize on it, as he advertised in the New York *Tribune* in March, 1860:

> P. T. Barnum, "the great American showman," . . . who furnishes more amusement for a quarter of a dollar than any other man in America, is . . . himself again. He has disposed of the last of those villainous clock notes, re-established his credit upon a cash basis, and once more comes forward to cater for the public amusement at the American Museum. Today . . . Mr. Barnum will appear upon his own stage, in his own costly character of the Yankee Clockmaker, for which he qualified himself, with the most reckless disregard of expense . . .

That same year the round-faced, professorial Chauncey Jerome gave his side of the story. Then in his sixties, he too had been wiped out and was too old to recover. As might be expected, Jerome believed his son had been hoaxed. In the first place he claimed that Terry and Barnum "cooked up a plan to sell their New York store (with its obsolete inventory) and the Bridgeport factory and machinery, if they could, to the Jerome Manufacturing Company, taking stock in that company for pay . . ." Barnum, he charged, grossly understated the liabilities of Terry and Barnum which Jerome had to assume: "The great difference in the real and supposed amount of their indebtedness and the unsaleable property turned in as stock were enough to ruin any company . . . if we had never seen Barnum we should still have been making clocks in that factory." A contributing factor to the disaster was the exorbitant interest rate which Jerome's son had to pay for bank credit. Jerome also refuted the showman's assertion that his endorsement of blank notes and drafts rendered him liable to a far greater extent than he realized.

Who duped whom? It is easier to believe the comparatively ingenuous clock manufacturer than the accomplished perpetrator of practical jokes. Henry Terry, the son of Eli, not only derided Jerome's skill as a clockmaker but questioned his business ability: "By misplaced confidence in other men, and by a disregard of rules of safety in pecuniary transactions, he was suddenly bereft of his estate." Likewise, Jerome's son seems to have exercised extremely poor business judgment. But Barnum was not so much a villain as a frightened novice investor operating in a field totally foreign to his experience. At first he tried to cover up the impending failure when the rumors began to circulate in December, 1855, but apparently none other than himself let the cat out of the bag. The Bridgeport *Standard* reported that the Jeromes hoped to survive without Barnum's aid through sales of their patented umbrella and corn-planter and the introduction of their new sewing machine. Instead, the Jerome Clock Company went bankrupt and, of course, never moved to East Bridgeport. Jerome's nephew, Hiram Camp, resurrected it as the New Haven Clock Company. Camp had begun making movements for his uncle in 1847 and boasted that "my taking the job of making Mr. Jerome's clocks . . . saved the clock business to New Haven people." Camp invented a machine for wire straightening and cutting that shaved production costs from thirty-five cents to seven cents per thousand pieces. At the start of the Civil War his company was disgorging 200,000 clocks a year, forty percent of the total then being produced in Connecticut. It takes a Yankee to fool a Yankee, and in the Barnum-Jerome fiasco each fooled the other.

*   *   *

Except for Terry Sr. and Joseph Ives, the clockmakers in Connecticut were predominantly manufacturing maestros. The cunning Jerome, for example, built his success entirely on the ideas of others. Nor were the younger Terrys, Seth Thomas or Hiram Camp inventors as such, but collectively they did create a great new industry before 1860. By that time most American-made clocks were of rolled brass, and most were produced by just four manufacturers: Seth Thomas of Plymouth; Elisha N. Welch, first president of Bristol Brass & Clock Company in 1850; William L. Gilbert of Winsted, a partner of Jerome until 1841; and the New Haven Clock Company. In one year these firms together accounted for over 800,000 brass clocks. Some retailed for as low as two dollars, a fraction of the prevailing price fifty years before.

Other than Jerome, the leading designers were Jonathan Brown, who achieved fame with his "Acorn" and "Rippled Gothic" shelf clocks from his Forestville factory, and Elias Ingraham. A native of Marlborough, Ingraham was apprenticed to a Glastonbury cabinetmaker and settled in Bristol in 1828 at the age of twenty-three. Bristol was rapidly becoming the clockmaking center not only of Connecticut but also the world. The end of the second war with England had encouraged the start of small enterprises, few of which made an entire timepiece. The clock fever was so widespread that many individuals simply bought parts and assembled clocks in their homes as a pastime. The historian J. W. Barber reported in 1836 that Bristol had sixteen clock factories, "in which

nearly 100,000 brass and wooden clocks have been manufactured in a single year."
Gideon Roberts's tall clocks of the craft era had given way to shelf clocks made
on machines with interchangeable parts.

Ingraham started his own clock case shop in 1831 but during the late
thirties, like so many other businessmen of that period, he ran into financial
trouble. Chauncey Ives lent him money; later he took a mortgage on Ingraham's
home and factory to cover the $8,000 loan; but Ingraham had to declare
bankruptcy. His brother Andrew stepped in and bought up everything at half its
value. The business continued, with Andrew watching the books and Elias
handling production and sales. From 1844 on the company made clocks with
brass movements, especially the popular "Sharp Gothic" shelf clock designed by
Elias. Spring-powered, this handsome clock was widely imitated. Like Jerome, the
Ingrahams found a market in England for cheap, one-day clocks.

As in gunmaking, it was customary for clockmakers to employ contractors.
Ingraham hired Anson Atwood to build some 12,000 spring-driven eight-day
brass movements and 5,000 time pieces. His contract stipulated that he must
"keep good order in the establishment, and allow no gambling, nor wrestling,
nor scuffing, nor profane language, have regular hours for business, and not allow
the factory to be opened on the Sabbath, except in the morning before Church,
and this only for the purpose of washing, shaving and preparing for church. He
and his hands shall be regular attendants at Church on the Sabbath. . . ."

Troubles of various kinds continued to plague E. & A. Ingraham &
Company right through the Civil War. Its shops were destroyed by fire in 1855,
although two-thirds of the $30,000 loss was insured. When business grew worse,
the brothers were saved by a mortgage on their property assumed by the Town of
Bristol.

As was true in the brass industry, the contributions of Connecticut
clockmakers centered on production. They added nothing to the principles of
keeping time. Weights and springs as the source of energy, and the pendulum as
the most practical kind of escapement, had been known for hundreds of years.
But in the quantity fabrication of clock parts, in the design of attractive cases and
in the conversion from wood to inexpensive rolled brass the Yankees excelled as
no one had before.

CHAPTER VII

Yankee Wares

THE peddling system of distribution, however infamous some of its practitioners, served to lay the foundations for the great metal industries in Connecticut—the brassmakers in Waterbury, the hardware manufacturers in New Britain and the silversmiths in Meriden. Thanks to the peddler, the burgeoning tin shops around Berlin after 1800 found an ever-widening market that gave the state a solid reputation for household wares. P. T. Barnum once said: "The Yankee peddler was the advertising medium for the tin shop and his wagon a trading post." Like most other early American craftsmen, the tinsmiths' work was primarily utilitarian rather than artistic, their products simple, prosaic and made for quick sale, yet today they are considered antiques of considerable value.

An exceptional tinware dealer was Oliver Filley of Bloomfield, who bought plain and japanned stock from Berlin tinmakers and in turn sold it to peddlers. He also started branch shops in New York and New Jersey. Filley made tinwares as well and had in his employ Oliver Bronson, "whose crooked-spout coffee-pots and pieced waiters placed the Filley products far above those of any other Connecticut tin-shop," as Shirley Spaulding DeVoe said in her study of *The Tinsmiths of Connecticut.* When his shop burned down in 1817, his workmen valued their jobs enough to contribute cash, labor and materials to rebuild it. Like as not, the tinsmith doubled as storekeeper, since he was usually paid in goods or farm produce, and his shop might be a converted house or barn. By no stretch of the imagination could he be called a manufacturer.

After learning the relatively simple art of tinmaking, Connecticut's pioneer metalworkers turned to pewter and Britannia ware, alloying tin with lead or antimony, and finally to brass and silver. The Yankee did not invent brass—the poor man's gold—nor was he the first to industrialize it; but with his customary input of ingenuity and zeal he did Americanize it. By 1830 two firms, with no more than thirty employees combined, had moved this industry from a household trade, beyond which tinmaking barely grew, to the factory stage. As the result of a bewildering succession of closely knit associations, job switchings and company name changes on the part of a few ambitious Yankees, eventually three large companies emerged in Waterbury to control the entire output of brass in the country and to supply a variety of household necessities, or luxuries made necessary. Except for waterpower, they had no natural advantage; their concentration in the isolated, inhospitable Naugatuck Valley was simply the fortuitous result of birthplace, individual initiative, excellent timing and early success.

Brass, of course, comes from melting and mixing copper and zinc, while copper holds the distinction of being, as far as we know, the first metal used by man. The "brass" mentioned in the Bible was either copper or bronze. Early Egytpians, with access to a rich vein of copper in Cyprus, combined it with tin to produce bronze and used the alloy for such purposes as rainwater drains, temple doors, swords and buttons. It took an Englishman, James Emerson, in 1781, to discover a process for making modern brass by the direct fusion of copper and zinc, and his countrymen soon mastered the casting, rolling and forming of the soft metal.

Around 1790 the three Grilley brothers, Henry, Silas and Samuel, somehow learned from an Englishman whom Henry had met in Boston how to cast pewter buttons in molds. This they did in their Waterbury home and added the bright idea of an iron eye to make them stronger and burr-free. In 1802, Abel and Levi Porter of Southington became associated with the Grilleys. Together they made the leap to rolled brass sheets, thus giving birth to the American brass industry—and, specifically, to the eventual formation of Scovill Manufacturing, the first of the Big Three. The Porters and Grilleys saw right in their own state an unfulfilled demand for pretty, glittering buttons to adorn the cottons and woolens being woven in almost every farmhouse. Tinware peddlers quickly added them to their packs. The entire work force, including the working partners, totaled thirteen. Over a pit fire superheated by a huge pair of bellows they melted scrap brass; then carried the ingots to a little iron mill in Bradleyville, near Litchfield, for rough rolling; after which the sheets were returned for finishing on two-inch steel rolls, operated by horse power, and for stamping and gilding. Output was limited to about ten pounds per hour. Until mid-century, copper and zinc had to be imported, since neither the meagre copper deposits in East Granby nor the collecting of old material, such as ship bottoms and utensils, could supply the demand.

To their distress the Porters and Grilleys found that a twenty-year headstart had enabled Great Britain to flood New England with buttons at a price way below the cost of the crude products from Waterbury. A new partnership, Leavenworth, Hayden & Scovill, succeeded them in 1811 and determined to strike back against foreign competition through improved methods. The only outsider of the original group, David Hayden, had learned buttonmaking in Attleboro, Massachusetts; he had also been a machinist for Sam Slater and operated his own textile mill. Dr. Frederick Leavenworth was both a physician and local storekeeper. Young James M. L. Scovill, the son of a Waterbury merchant and large landholder, had clerked in his father's store. In 1817 Hayden designed a brass lamp and built a machine to cover buttons with cloth, a distinctive product feature for Scovill to boast about on his frequent selling trips. Scovill also handled the purchasing of raw materials.

One of their best decisions was the hiring of the first experienced gilt buttonmaker to come from England, James Croft, who taught them how to anneal copper and zinc to produce the correct orange tint that made brass so popular and so adaptable to thousands of household uses, big and small. But Croft soon defected to a competitor, Aaron Benedict, who encouraged him to sneak back to his homeland and return with the latest in rolling equipment to make sheet brass, thus converting the embryonic brass industry from a household to a truly factory operation.

After the first decade, none of the five original pioneers remained. The second generation—the Scovills, Aaron Benedict, Israel Holmes, Israel Coe and Anson Phelps—were mainly responsible for the entrenchment of the new industry. Benedict had hoped to follow his older brother through Yale, but family illness in his sophomore year ended his college career and forced him into retailing, a trade he intensely disliked. His initial stroke of luck was to marry the daughter of Abel Porter, the first brass buttonmaker. In 1812, Benedict decided that manufacturing bone and ivory buttons might be the quickest route to fortune; upon the declaration of war against England, however, he shifted to pewter buttons for the army and navy. A decade later he took in four partners, including James Croft, who put up a total of $6,500 and joined the brass button bandwagon.

It was in 1824 that Croft journeyed to England for a pair of steel rolls, thirty inches long by eleven inches in diameter, the largest yet seen in America. Sheet brass is produced by melting copper and zinc in roughly a two-to-one ratio, casting the alloy into an ingot and then passing it repeatedly through rolls until the desired thickness is obtained. Now Benedict could supply all the brass he needed for his own use and have a surplus to sell to others, like the New Britain hardware men and clockmakers in the Bristol area. In New Haven, Chauncey Jerome used it for his newly invented one-day clock, and brass clocks became America's biggest export made from this metal. Besides buttons, Benedict produced butt hinges and, after seizing upon the English discovery of electro-plating in the early forties, bolts, hooks and eyes, rods, tubing wire, pins and finally whale oil lamps. By 1840 the firm of Benedict & Burnham had assets of $100,000 and became the first brass factory in Waterbury to incorporate. Later Benedict spun off two of his divisions into separate companies—Waterbury Button and Waterbury Clock. In 1900, the parent formed the nucleus of today's American Brass Company, the second and largest of the Big Three.

In the meantime, Scovill with his brother William had bought out Leavenworth and Hayden in 1827 for $10,000. They followed Benedict's lead in converting to brass stampings and, thanks to a wily employee named Israel Holmes, went him one step better. A native of Waterbury with a smattering of experience teaching school and selling hats, Holmes had married a Bristol Barnes. On a trip to England in 1829 to obtain brass machinery, he persuaded twenty British journeymen, in defiance of the crown, to accompany him back to America, concealing them in empty wine casks until the ship left Liverpool. The same year the button shop burned to the ground but was immediately rebuilt. For the next quarter century, the Scovills worked in brotherly harmony, adding such products as safety pins, coins and tokens, needles and lamps. Their personalities were complementary: James, the older, being the impulsive optimist always ready to try something new, while the more prudent and painstaking William ran the plant, kept the books and negotiated bank loans. After the introduction of Daguerre's photographic process in France, they produced the first silver-coated copper plates for the soon popular daguerreotypes (1839); they also supplied nickel silver to the Meriden silversmiths. Another Scovill employee, Hiram W. Hayden, invented a method for forming brass kettles by spinning.

Scovill buttons were at first transported by wagon or stagecoach to New Haven for shipment to New York. To improve the service, the Scovills subsidized the stagecoach company and later bought stock in the Naugatuck Railroad, promoted by Israel Holmes, which ran to Bridgeport. Occasionally they stocked peddlers who paid up only after selling the merchandise. For example, in 1828 Jason Atwood and George Merriam were hired for thirty-five dollars a month each to go to Ohio. Cash was the biggest headache. From Ohio Atwood wrote: "I swapt the thread for books and cotton cloth and Broadcloth and I think we can turn them into horses and cash but we cannot sell the buttons." He finally solved the dilemma by taking in trade 395 gallons of whiskey and a half dozen horses and bringing them back to Waterbury for sale. But selling through agents or merchants in places like Boston, New York or Philadelphia proved more lucrative, although the extending of credit constantly bedeviled the brothers. Promissory notes for raw materials, store goods or labor performed frequently ran ninety days or longer; when banking became available, the drafts on agents or merchants could be discounted. The panic of '37 dried up bank credit, forcing the Scovills to lay off workmen and to liquidate some of their assets at a loss. "At least," observed William, "we are about as well off as our neighbors . . . If we can squeeze through without stopping shall be thankful and will live without spending money until we get our debts reduced." At that time they were in debt about $70,000, yet managed to show a profit.

Sales grew steadily, rising from $10,000 in 1823 to $35,000 by 1830, equivalent to an output of 100 gross or more of buttons per day. The variety also multiplied. By 1832 Waterbury had three gilt button shops, causing James to comment to his brother: "The Business has been brought down to a complete Tin Peddling, Jockeying concern and an honorable and fair competition is out of the question." This situation stimulated their quest for new ventures, such as plated silver and sheet brass and fabricated articles like hinges and daguerreotype plates. Even so, the profit margin on buttons by today's standards would have made the modern manufacturer envious; labor and raw material cost no more than twenty-eight cents a pound, while the selling price was seventy-five cents per pound.

The rolling of sheet brass, they found, required lots of waterpower. Two waterwheels supplied power for the rolling mill until the Scovills bought, in 1852, a steam engine for $4,000. Back in 1830 they had investigated the possibility of using steam, perhaps looking at Hartford's Woodruff & Beach engine which burned a cord of wood a day. Instead, they acquired another water privilege and erected a breast wheel. But the frequent necessity to shut down operations because of the lack of water during a summer dry spell or winter freeze convinced them of steam's superiority. In winter all hands would too often have to be sent outdoors to cut the thick ice from the dam in order to keep water flowing.

The Scovills' conversion to steam power coincided with the abandonment of their general store, an essential feature of almost every early manufacturing community when wages were usually paid in credit and on a yearly basis. For several years, however, the Scovills experimented with the weekly payment of wages in cash until the depression of 1837 killed such an idea. James then advised his brother: "As to paying the hands I would not do it at this time if they do not like our Notes let them have Buttons or Notes we have agst. other People . . ." Near mid-century a savings bank opened in Waterbury, accelerating the urbanization of the factory worker. The Scovill store carried a wide assortment of goods, many of which were taken in trade from their customers, and others, like hats and shoes, made by local artisans.

Five years passed after Croft's successful expedition to England for rolling equipment before the Scovills could hire their own expert casters. Workers from abroad, like those induced to emigrate by Holmes, usually signed five-year contracts; they adapted quickly to the land of liberty and because of their expertise acted quite independently. In a doleful New Year's Day letter James, harassed by the breakdown of the new equipment, told his brother: "Our English hands are all keeping New Years this day & of course will not work tomorrow Fairclough & all." But the partners did not dare to discharge any of them, no matter how uncooperative they might be. Their wages ranged from $500 to $1,400 a year; they guarded their skills so jealously that the Scovills tried subcontracting by piecework, thereby increasing take-home pay, in order to encourage them to train native apprentices. The daily hours of work varied from twelve in summer to nine or ten in winter. Women were employed at $2.50 weekly to cramp, clean and pack the buttons. At mid-century the Scovills incorporated with capital of $200,000 and 190 employees, for two decades having enjoyed the advantage of being the only brass rolling factory and now rapidly emerging as the largest organization of its kind in the United States.

Tempted by the ample profit margins which Benedict and the Scovills piled up, in some years 100 percent or more, Israel Holmes left the Scovills early in the game and raised $8,000 to manufacture brass on his own. The next year, 1831, he sailed again to Birmingham to filch a wire drawer and more artisans willing to emigrate. Over-capacity of both sheet and wire brass now forced the eager brassmakers into fabricating consumer products. In 1838, the partnership of Holmes & Hotchkiss changed its name to Brown & Elton; they introduced brazed brass tubing, sold their surplus of brass wire to other manufacturers, and made hooks and eyes. They also entered the pin business, buying out the Fowlers of Northford, who in competition with the Howe Manufacturing Company and Slocum & Jillson in Poughkeepsie, New York, had pioneered this specialty.

Articles Carried at the Scovill General Store in Waterbury

(SOURCE: J. M. L. and W. H. Scovill *Inventory Book of 1839–1845*.)

Cloth (all varieties)	Umbrellas	Twine	Locks
Blankets (incl. horse)	Straw hats	Ink	Watchkeys
Skirts	Fans	Sand boxes	Spoons
Table covers	Combs	Pens	Scissors
Vesting	Boots	Stamps	Guns
Drawers	Thread	Pencils (silver)	Chisels
Wrappers	Buttons	Hooks & Eyes	Knives & forks
Shawls	Toilet bottles	Paint	Shoe Knives
Silk	Soap boxes	Indigo	Saw sets
Linen	Soap	Locofocos	Plyers
Cravats	Lip salve	Rope	Stove
Handkerchiefs	Brushes	Glue	Springs
Superbas	Thimbles	Candles	Saws
Oil cloth	Toothbrushes		Razors

Hose (incl. silk)	Bodkins		Screws
Yarn	Needles		Iron butts
Lace	Wallets		Brass butts
Aprons	Pins		Spectacles
Bags	Gloves		Spectacle cases
Tapes	Baskets		Trowels
Veils	Hats		Brace
Collars	Paper boxes		Skates
Ribbons			Casters
	Razor strap	Groceries	Door latch
Dishes	Whips	Sal Eratus	Oil stove
Cups	Boats	Tobacco	Bolts
Clock			Chest locks
Teapot	Pots		Latches
Coffee mills	Stoneware		Gimblets
Lamp	Looking glasses		Stirrups
Lanthorn	Trays (steel)		Scales
Candlestick			Pocket Knives
Cannister	Stools		Tacks
Gridiron	Door Shutters		Brads
Shovel & tongs	Chairs		Files
Fire dogs	Trunk		Scythe
Toaster	Broom	1 lot burnishing	Hoes
Kettles	Tubs	stones $50.00	Seives

Strange to say, the common pin used to be a relatively expensive item for the lady of the house. A package of pins in 1812 cost one dollar. The phrase "pin money" derives from the allowance given by thrifty colonial husbands to their wives for acquiring these essential articles. They were imported from England until a New Yorker, Dr. John Ireland Howe, invented a machine in 1831 to make a pin in one operation—an amazing feat when one realizes that prior to that time pins had to be hand formed in eighteen operations, one worker being able to finish only 480 an hour. Howe subsequently set up a plant in Derby. Meanwhile the Fowler brothers also invented a similar and superior machine. But when Dr. Howe introduced the solid-head pin and perfected a device for sticking the little pins into paper packets, his ingenuity won the pin game. In 1846, the astute Israel Holmes teamed up with Benedict on a fifty-fifty basis to combine the three pinmakers into the American Pin Company.

Thomas Low Nichols wrote a vivid account of this early example of automation:

Many years ago I visited the village of *Waterbury*, in Connecticut, and spent a day among its curious factories. Water and Steam power were at work, but comparatively few human operatives. In one large room, full of machinery in rapid motion, there was but one man, whose business was to watch the machines and supply them with material. Each machine had a great coil of *brass wire* on a reel beside it. The end of the wire was placed in the machine, and from it flowed hooks or eyes into a basket as fast as one could count. These machines required only to be fed with coils of wire as they were used. In another room, automatic machines were cutting up coils of iron-wire and discharging hair pins. Brass-wire went into other machines and came out common pins, with heads and points all perfect, and only requiring to be tinned and papered. The papering was done by a machine which picked out the pins, laid them in rows, and then pushed each row into a paper. One pin factory made three hundred thousand dozens of pins a day. Another machine took wire from a coil and bits of brass from a hopper, and turned out buttons with the eyes made, set, and riveted.

For some years Yankee waremakers were somewhat protected by a levy on imports, but in 1833 Israel Holmes and Israel Coe were chosen to represent the Naugatuck Valley manufacturers in Washington to protest a bill that would classify sheet brass and wire as nonmanufactured items to be admitted free of duty. They persuaded the "Great Compromiser" Henry Clay to draft and push through Congress a separate law that maintained the tariff rate established in 1818, giving the industry an advantage over its English competition until the Civil War. By that time, not only did brassmakers have plenty of skilled workmen trained, but their know-how surpassed that of the English, and Waterbury interests dominated the nation's domestic output.

When Holmes returned home after his political victory, he met with a personal tragedy which nearly terminated his business career. During his absence his home had burned and two of his children perished. He retired from Holmes & Hotchkiss and went into seclusion for a while. But this tight-fisted go-getter and super salesman could not remain idle long. Before his death Holmes became Waterbury's leading industrialist, organizing five different companies. Wolcott-ville Brass made kettles, the loud hammering of which was a great irritant to the quiet-loving residents of that village. Waterbury Brass (1845), the first to follow Scovill in rolling brass, was the largest mill yet built and an important supplier of brass and nickel silver, sheet copper, wire, rods and tubes, as well as a manufacturer of rivets, jack chains and lamps. Bristol Brass & Clock (1850) made clocks for a decade but after taking over the assets of the floundering Wolcottville Brass Company during the depression of 1857 concentrated on sheet brass. In 1852 Waterbury Brass bought Hiram Hayden's spinning process from the Scovills and soon controlled the brass kettle market. The following year Holmes went into competition with Scovill on daguerreotype plates, hiring a former employee of the French inventor Daguerre. After the Civil War he was instrumental in founding Plume & Atwood, whose chief product was the Rochester oil lamp.

There were others who "wanted in" on the brass boom. Sales-minded Israel Coe, one of Benedict's partners, broke away in 1834 to join Anson G. Phelps of New York, and Holmes in the Wolcottville Brass Company. Coe also was the moving force in organizing his fellow brassmakers to exploit the rich copper deposits around Lake Superior by building the first smelter in Detroit. The only outsider to enter the industry, except for David Hayden, Phelps was an importer of tin, brass and copper but yearned to partake of the gold that his Waterbury friends were finding in brass. In 1836 he built a copper rolling mill in Derby. Seized by a vision of making the new settlement a thriving factory town, he named it Birmingham, after the English city, and attempted to buy up all the surrounding land. A local farmer thwarted his scheme by acquiring a key piece of property and escalating the price beyond reason. Backing out, Phelps then moved two miles up the Naugatuck River and, in 1844, organized Ansonia Brass, at the same time founding the town of Ansonia and giving it his Christian name. Today, Ansonia Brass & Copper belongs to the American Brass complex.

Another Waterbury native, Almon Farrel, a millwright by trade, and his son Franklin established in Ansonia what became the biggest producer of heavy castings and large-scale mill machinery for iron, brass and copper. They came there for the purpose of surveying and supervising the construction of Phelps's canal and copper mill, and in 1848 built a small foundry and machine shop on land acquired from Phelps, including "one-half square foot of permanent water." Soon they purchased the Waterbury Iron Foundry. Their first products were brass and iron castings and power drives and gears for water power installations, from which they branched out into rolling mills for the brass companies in Waterbury and calenders for the rubber companies in Naugatuck. From the beginning the Farrels were sticklers for high-quality workmanship.

The third of the Big Three, Chase Brass & Copper, did not appear until World War I; one of its affiliates, Waterbury Manufacturing had, in turn, acquired the old U.S. Button Company, the predecessor of which began in 1837.

The buttons of bygone days that spewed from Waterbury reflect the history of their time, bringing back to life the men who wore Union Blue or Confederate Gray; milady's tight bodice buttoned from stem to stern; the coin cuff links worn by gentlemen, along with stick pins bearing a state seal; police badges, fraternity and society emblems; political campaign buttons; brass milk bottle covers, huge belt buckles, brass trimming for bugles, candy tongs and parchesi counters. When General Lafayette revisited America in 1824, the Scovills presented him with a set of solid gold buttons bearing an embossed profile of George Washington. Supporters of Andrew Jackson during his presidential campaign in 1828 ordered metal buttons identifying themselves as "Old Rough and Ready" diehards. President William Henry Harrison revealed his political know-how in his 1848 campaign by having buttons struck off which showed a cider barrel at the front door of a log cabin, the latter symbolic of his birthplace, to please those who favored hard cider. For those sections where drink was anathema, the cider cask was omitted. When the widespread use of paper money during the 1830s put a premium on specie, Scovill medals circulated as currency. Some of their copper coins bore patriotic inscriptions like "not one cent for tribute." The U.S. District Court acted to stop the issuing of tokens, in effect a private mint, yet Scovill continued to make them, being careful to avoid legal sanctions by omitting the reproduction of a human head on one side and an eagle on the other.

* * *

Paralleling the growth of brass in the Naugatuck Valley was that of hardware in Bristol, Collinsville, Southington and particularly New Britain. For 200 years the fire, bellows, anvil and hammer of the village blacksmith took care of the latches, hinges, bolts and locks needed for every new farmhouse, hardware for horses and wagons and farm implements. Soon after the Pattisons started banging out tinware in Berlin, families with such indigenous names as Hart, Judd, North and Stanley operated smithies in the western section of that town, which became New Britain in 1850. Blacksmith shops now became machine shops. Here a descendant of a first settler, James North, hammered out an endless variety of tools—augers, brads, bridle-bits, chest locks, nails, cranks, chisels, crowbars, hooks, knobs, keys and spikes, many of which the peddler carried away on horseback.

Around 1800 North sent his oldest son James to Stockbridge, Massachusetts, along with two other young men named Joseph Booth and Joseph Shipman, to learn how to cast brass. Each of the three set up a foundry, from which tinkling sleigh bells soon enhanced the allure of the peddler's load. James turned his business over to his younger brothers Seth, Alvin, Henry and William. In 1812 Seth and Alvin formed a partnership for making buckles out of plated wire imported from Europe. The wire had to be drawn, cut and shaped by hand, labor that the Norths subcontracted to women working at home. Their company, under the name of North & Judd (1855), eventually grew into the largest manufacturer of saddlery hardware in the country. William North branched out into jewelry. In 1830 Levi Lincoln of Hartford built for Henry North the first machine for hooks and eyes, enabling the Norths to compete with Israel Holmes and the Scovills in Waterbury.

Thus, at the same time that Waterbury and Meriden were taking the lead, respectively, in brass and silver, New Britain was well on the way to earning its nickname of the "Hardware City." The *Greenfield Gazette* reported in 1836 that "the village of New Britain . . . is as famous for the manufacture of *nick nacks* as

Scovill Manufacturing, 1858

Stanley's Bolt Manufactory, 1843

any other portion of the land of steady habits . . . There is $1,300,000 worth of little indispensibles—such as stocks, suspenders, brass castings of all kinds, etc. turned out there annually . . . The suspender factory employs 200 persons . . ."

Just as a triumvirate in Waterbury dominated the brass industry, so three firms in New Britain accounted for most of the builders' hardware made in America, the development occurring mainly after 1837 and culminating in joint stock companies after 1850. Two of the pioneering firms, Russell & Erwin and P. & F. Corbin, later merged into American Hardware. Cornelius B. Erwin, a native of Booneville, New York, came to New Britain as a drover with a consignment of horses. For a while he worked for James and Seth North, and in 1836 married Maria North, James's daughter. In 1839, he formed a partnership with Henry E. Russell, and they took over the plate lock factory founded earlier by Frederick T. Stanley. Russell & Erwin (incorporated 1851) was the first to specialize in builders' hardware.

At age fifteen, Philip Corbin's ambition to forsake his farm in West Hartford was fired by gossip of the liberal wages to be earned in a New Britain factory, much more than the fifteen dollars a month he could get from chopping wood or tilling a field. In 1844, he went to work at Russell & Erwin as an apprentice to one of the lock contractors or foremen. To his chagrin the pay was a dollar less than on the farm; to make up the difference, he took on odd jobs after his regular work, including sweeping out the entire factory every week for fifty cents. The next year he got a better-paying job with North & Stanley, another lockmaker. By dint of working nights to learn the business thoroughly under the sharp eye of Stephen Bucknell, the first cabinet lockmaker in America, who hailed from Watertown, Corbin himself was accepted as a plate lock contractor at twenty, the youngest ever. Five years later he was ready for a partnership with his brother and a brassmaker, each of whom invested the meagre sum of $300. That, however, took almost every nickel Philip Corbin had saved; he held back $18.00 which he concealed in his bureau drawer and asked his wife to spend only when absolutely necessary. She made it last twenty months. Despite the fact that imported hardware was still preferred by discriminating customers, the Corbins prospered in the old North & Stanley plant. Three more brothers joined them. Their first product was brass balls for tipping the horns of oxen, surely one of the most unique articles ever manufactured of that material. They competed with Scovill on wrought brass butts, adding to their line such intriguing and first-time items as flush bolts and coffin fittings, and by selling through both dealers and their own salesmen became the largest of the independent hardware firms.

Stanley Works had its origin in the pioneering of Frederick T. Stanley, whose family came from Farmington. When Stanley started making iron bolts

Frederick T. Stanley. Founder and president of the Stanley Works, 1843–83

for doors by hand in a nondescript one-story building in 1843, New Britain, like Waterbury, was still an isolated community of about 3,000, connected to Hartford only by a stagecoach that ran thrice weekly and cost a quarter to ride. The new railroad between New Haven and Hartford lay two miles to the east. The unusual thing about Stanley's operation was the single-cylinder, high-pressure steam engine he had bought in Brooklyn and transported up the Connecticut River and then by oxcart from Middletown. Like so many other early manufacturers, he filled every position from production manager to door-to-door salesman. As did Charles Parker in Meriden, he also exhibited a strong sense of civic responsibility by becoming New Britain's first mayor. After incorporation in 1852, butt hinges were added to the line and the work force increased to twenty-five men. One of the incorporators was J. M. L. Scovill, founder of Waterbury's Scovill Manufacturing Company, illustrating the interconnections that were common between different industries and their leaders.

The real growth of the company, however, coincided with the employment in 1854 of William H. Hart, at the age of nineteen, as secretary and treasurer. A dynamo who held the business together during the trying period following the panic of '57 and for a span of sixty-three years, thirty-two of them as president, Hart steered Stanley's progress with unrelenting determination. He was an organizer of first-rate ability, a man of uncommon vision and a fierce competitor. According to one story he would wake up at night with an idea, and to make sure he remembered it, tie a knot in his handkerchief and throw it into the middle of his bedroom floor. Frugal to the nth degree, he had the office boy save envelopes from incoming mail for his scratch paper. He pioneered the cold rolling of steel, the immediate result being a uniform door hinge.

The third member of the hardware Big Three, Landers, Frary & Clark, which specialized in table cutlery, began as a partnership in 1842. Like Erwin an outsider, George M. Landers continued alone after 1847, making casters, cupboard catches, coat and hat hooks and other small articles. Connecticut's first cutlery company was Pratt, Ropes & Webb in Meriden (1845). The Pratt in this firm was Julius, who achieved fame as the dean of the ivory comb industry. Ivory handles soon gave way to wood and, following Goodyear's invention of vulcanization, hard rubber. Another pioneer was Peck, Stowe & Wilcox in Southington, also a tinware center. Seth Peck saw a need for a practical tin-making machine and in collaboration with his brother-in-law, Edward M. Converse, obtained a patent in 1819. Converse seems to deserve the major share of the credit for the invention, and in the next few years perfected a complete set of tinners' machines, which used steel rolls and the swaging principle. But he sold out his interest in 1832, and Peck reaped the harvest by manufacturing what

William H. Hart, 1834–1919

Converse had created. As has so often been the case in industry, the inventor shook the bush and the capitalist caught the bird.

In 1847 the life of Middletown still revolved around shipping. Shipyards, warehouses and sail lofts thrived. One cold winter's day a group of idlers warmed their toes before the stove of Ben Butler's loft on the river bank; a foreman was trying to thaw out a frozen topsail in order to replace the worn-out rope grommets. "If some of you confounded loafers would get to work and invent a metal grommet," he complained, "we wouldn't have to do a mean job like this." To Eldridge Penfield, a young clerk, this was a challenge not to be ignored, and he soon developed and patented a metal grommet for fastening sails. He interested his uncle in a partnership, and with the help of a friend named William W. Wilcox, they stamped out the tiny article on hand presses. When the inventor grew frustrated over the slow progress of the business, he sold out to his uncle and to Wilcox, who had saved $250 from his meagre pay of 75¢ a day. Wisely, Wilcox got rid of their greedy agents and set out to sell the grommets directly to sail lofts along the coast. Soon he added other items for the sailmaker like thimbles, spectacle clews, rings and sticking tommies. After ten years, Penfield & Wilcox dissolved, upon Penfield's retirement, and Wilcox continued on his own, eventually forming Wilcox, Crittenden & Company, a still flourishing marine hardware enterprise.

*   *   *

In colonial times, especially around Boston, silversmiths like Paul Revere and William Moulton abounded as one-man shops dependent upon coins for their raw material. They made everything to order, and embellished the dining rooms of wealthy merchants with tankards, porringers, candlesticks and spice boxes, as well as cutlery. Silver buttons and knee buckles were the mark of a gentleman. But Connecticut transformed silverware into an industry that eventually accounted for over half of the total production in America.

Before the advent of silverware, small tin, pewter and Britannia shops flourished around Meriden in the 1830s. Ashbil Griswold and the Curtises were especially noted for their high-grade pewter kitchen and tablewares, such as plates, platters, basins, mugs and spoons. The Yales, tin peddlers who gave their name to Yalesville in Wallingford, were the largest manufacturers of pewter and Britannia hollowware—tea sets and communion services in particular. The prolific Danforth family in both Norwich and Rocky Hill had been putting their mark on pewter since 1733. Thomas Danforth III trained Charles and Hiram Yale and also Griswold. Usually, one part tin and three parts lead (sometimes brass or copper), pewter proved a readily accessible substitute when pure tin was unobtainable. It could be easily melted and reused. Hollow-ware pewter, for pots

Wilcox, Crittenden & Company, Middletown

Meriden Britannia Company

and tankards, contained 20 percent lead, while flatware pewter used antimony instead of copper. A pewterer had to make a large investment in molds, soldering irons, other tools and materials. Unlike Sheffield plate, which the wealthy preferred, it could not be polished and required an inordinate amount of scouring. Pewterers in the Connecticut Valley accounted for three-quarters of all colonial tableware.

Ashbil Griswold started in business in 1808, making mugs, spoons and pots; a leading citizen of Meriden, he served in the legislature five times and was also president of the town's first bank. By the time he retired in 1842, Britannia was on its way to replacing pewter. An alloy of tin with small proportions of antimony and copper, it was much harder, more wear-resistant and, unlike pewter, could be polished to a high lustre.

For the transition to silverware, Robert Wallace in nearby Wallingford deserves credit as the pioneer. At the age of sixteen he had been apprenticed to Captain William Mix of Prospect to learn the art of making Britannia spoons. Two years later, in 1833, despite his youth, he started his own business in an old gristmill which he leased in Cheshire. One day a customer in New Haven showed him a spoon he had just received from England made of a strange new metal. Wallace readily saw that it was tougher and brighter than pewter, even harder than Britannia, and he wasted no time going to New York for a bar of it. Soon he was the first in this country to produce a flatware of German silver, locating his operation on the Quinnipiac River. German silver, actually an alloy of copper, nickel and zinc, had the additional virtue of being adaptable for rolling. For better control of his material, Wallace found a man in Waterbury who knew the exact formula and was willing to part with it for twenty-five dollars. For twenty years he supplied spoons under contract to Deacon Almer Hall, another local silverware manufacturer, and others who were more adept than he at disposing of them. Later he formed a partnership with Samuel Simpson under the name of R. Wallace & Company; upon its incorporation in 1865, Meriden Britannia acquired a one-third interest.

The Rogers boys in Hartford, however, were in the meantime making a far more significant contribution to the growth of the silverware industry than Wallace. William, one of twelve children, was apprenticed in 1820 to a jeweler and silversmith named Joseph Church, who made fine spoons out of silver coin for the carriage trade. Five years later Rogers had advanced to partner and brought in his brothers, Asa and Simeon, to help. By 1836 the Rogers Brothers were running their own enterprise, dealing in "spoons and forks and soup ladles as pure as coin."

Asa apparently wearied of being a silver artisan, sold out to William and moved near the Tariffville carpet factory. Here, in association with William B.

W. Wilcox & Company Lock Factory, Middletown

Hall, Elton & Company, Wallingford, 1828, spoon manufacturers. Robert Wallace, first to use German silver, was associated with Hall until 1855

Cowles, a spoon manufacturer, and James Isaacson, he began in 1843 to experiment with the brand new English process of electroplating. How this invention only three years earlier could have been so quickly purloined and brought to the attention of East Granby spoonmakers as well as Waterbury brassmakers is still a minor mystery. Asa learned the miraculous art of applying, by induction, a thin coating of silver evenly and firmly to a base of copper, nickel and zinc (or German silver). The trick was to use a cyanide solution, with the copper and zinc acting as an electric conductor, causing the pure silver to dissolve and reform upon the surface of the article being plated. Eureka! Asa must have felt that he held the key to the mass production of silverware that would not only render obsolete expensive Sheffield plate but enable every farmer and millworker to buy it. But discovery was one thing, production another. Cowles Manufacturing did convert from Britannia to silver-plated forks and spoons, and the name "Spoonville" for that section of East Granby became almost as well known as Tariffville. At the end of August, 1846, however, Isaacson wrote William Rogers a letter of abject distress concerning the company's financial plight and the poor workmanship. The next day Rogers took a mortgage for $5,150 on the Cowles factory. Late that year Asa quietly dropped out of this sad situation and returned with his electroplating know-how to his brothers.

The following March the Hartford *Courant* reported that "the Rogers Brothers Store which has been located on State Street and has been headquarters for cutlery as well as making a specialty of silverware made exclusively of dollars are now producing an entirely new and novel article in silverware . . . They are importing German silver spoons and forks, which by a new and unique process are coated with pure silver." That year, 1847, they sold more than $20,000 worth of silver-plated articles. The *Courant*'s "Man About Town" enthused over their work, as he watched them replate Senator John C. Calhoun's family tea urn:

> The plating you see is perfectly compact and solid, and of any thickness you wish . . . What beautiful sets of plated communion-service, tea service, and hollow-ware are these around us . . . Fifty or more hands are employed in burnishing, finishing and packing the work . . .

In the late fifties the three brothers operated plants in both Hartford and Waterbury.

Meanwhile tinware's lead in Connecticut metal-working shops was beginning to fade before the onrush of Britannia and brass wares and the output of clocks. Meriden was a beehive of manufacturing activity, with nearly a quarter of its population of 3,000 employed in thirty-five factories. The Yales and nine other firms still poured out a variety of wares, but the chief industry was, surprisingly, ivory combs. Julius Pratt imported elephant tusks from Africa weighing up to eighty pounds apiece, from which he stamped out combs in a series of machine operations. At one time three of every four ivory combs in America, over 15,000 per day, came from the Pratt factory, which during the Civil War moved to Deep River under the name of Pratt, Reed & Company.

Then on the scene appeared two unusually keen and energetic young peddlers, Horace and Dennis Wilcox, from a family almost as large as the Rogers's. Their brother Jedediah became known for his carpet bags, to which he added the manufacture of corsets and hoopskirts. On local peddling trips, Horace had listened hard to customers' complaints about the drawbacks of his wares—tin kettles and pots whose ears and handles fell off, pewter that was too hard to clean, Britannia whose shine didn't last. He knew of the Rogers Brothers and their reputation for silverware, made their acquaintance and began to carry their silver-plated spoons, forks, cups and candlesticks with instant success. He found they outsold every other kind of ware.

From his experience in selling the various products of several small shops, Horace Wilcox suddenly had an inspiration of an organizational kind. Most of the shops he represented were woefully undercapitalized; the average for all manufacturing in the United States was only $25,000. Distribution likewise posed a major problem. Accordingly, in 1852, he founded the Meriden Britannia Company, a seven-man partnership, all of whom either made or sold pewter and Britannia wares and had combined annual sales of $50,000. His intention to have the company be primarily a selling organization soon broadened to include manufacturing. The traditional casting of articles in molds gave way to the rolling, spinning and stamping of nickel silver, which Wallace had introduced, and then to electroplating. At first, electroplated wares met with resistance by quality-minded consumers due to the popularity of imported Sheffield plate, but soon even the most skeptical and fussiest buyers appreciated their beauty and durability. The Wilcoxes' first catalog featured a broad list of items including silver-plated spittoons for $3.75 each, shaving boxes, liquor mixers, toys and japanned tinware. Their sales in 1853 totaled $250,000, half of which were products made by other than Meriden Britannia. Later the company offered plated hollow-ware and flatware, as well as 104 different dinner casters and twelve ice pitchers. Gradually Britannia ware disappeared entirely.

By 1860 the Wilcoxes' annual sales had climbed to half a million and 320 were employed. When the Rogers family suffered financial reverses in 1862, Wilcox seized his opportunity, bought their equipment and services, built a large new factory, and from then on made Meriden the home of the most famous name in silverplate. William Rogers, however, did not relish working under any one else; although still under contract, he returned to Hartford and with his son operated independently until his death in 1873. By the turn of the century all of the original silver firms had been consolidated into the great International Silver Company, which continued to use the "1847 Rogers Brothers" trademark.

While the Waterbury brassmakers, the New Britain hardware manufacturers and the Meriden silverworkers were flourishing, other important industries took root in Connecticut's fertile manufacturing soil, such as the carriage business in New Haven, the hat industry in Danbury, the Cheney silk mills in South Manchester, and the great Colt Armory in Hartford. Their contributions are covered in the next chapter.

Brewster Carriage Factory, New Haven

George T. Newhall's Carriage Emporium, New Haven

CHAPTER VIII

Success and Failure

IT would be next to impossible to account for all of the enterprises founded in Connecticut before the Civil War in pursuit of the personal calling which Cotton Mather had urged every good Christian to undertake. Many, of course, were failures—some in a very few years, others after decades. Companies, like people, die for a variety of reasons. Numerous fine old Yankee firms have passed from the scene due to obsolescence of their product, the atrophying of management, or—in a last ditch effort at salvation—merger. The old saw about riches to rags in three generations applies as often to companies as to families. Names of once-thriving manufacturers like Cheney Brothers, Collins, Weed Sewing Machine, Pratt & Cady, Pope-Hartford and Billings & Spencer are only memories today.

Occasionally, an entire industry succumbed to the relentless march of technology, as in the case of carriages. Carriage building by chance centered in New Haven. In fact, that city made more carriages than any other in New England. In the year 1809, as it passed through New Haven, the stage carrying a young man from Preston broke down. Detained against his will, James Brewster wandered around to see the sights and happened to meet John Cook, the first builder of gigs in Connecticut. The conversation resulted in Cook offering Brewster a job. A year later, having saved $250 from his salary, Brewster married and started his own shop.

He was no novice, having learned the wagonmaker's trade in Northampton. He also had strong moral convictions, being a devotee of Benjamin Franklin. It is said that he often studied Franklin's maxims an hour a day, even though he used to work twelve hours, six days a week. Following Franklin's commandment that a trade was better than a profession, Brewster claimed he had rejected an offer of a college education.

Aware that the quality of Cook's carriages left much room for improvement, he tried a different tack in order to compete with the best English equipages. The most popular vehicle of that time was the one-horse "buggy." Eschewing this or the runabout, he went in for high-quality phaetons, victorias and coaches that appealed especially to affluent Southerners. By 1827, he was big enough to have a warehouse and repair shop in New York. Soon his carriages were being exported to Mexico, South America and Cuba. In 1832, he had the honor of making special carriages for President Jackson and Vice-President Van Buren.

Brewster was a fair-minded but severe employer who, like John Wilkinson and Samuel Collins, would not stand for the use of liquor on the job, a practice which many shops seemed to tolerate. By offering exceptional wages and paying cash every Saturday night, he attracted a superior group of artisans. In what amounted to a startling fringe benefit for that period, their training was broadened by special evening lectures and scientific courses given by Yale professors. Among those whom Brewster engaged to instruct his workers at his "Young Men's Mechanical Institute" was Professor Benjamin Silliman, the gifted teacher and scientist who edited the *American Journal of Science and Arts*.

Brewster also took a keen interest in the welfare of his town. He planted 300 of New Haven's once-famous elm trees; he was a benefactor to the poor and to orphans. He was the chief promoter of the railroad to Hartford and served as its first president. Although his fortune was depleted by severe losses in the railroad venture, he retired in comfortable circumstances in 1856.

A bevy of competitors followed in his footsteps, branching out into carriage hardware, axles and springs. The elliptical steel spring had been invented by Jonathan Mix in New Haven back in 1807. Chauncey Jerome recalled that he lived in the immediate vicinity of the largest carriage maker in the world (probably George T. Newhall's shop), which "turns out a finished carriage every hour!—much of the work being done by machinery and systemized in much the same manner as clockmaking."

An Englishman who visited New Haven briefly in 1848 noted the "numerous small manufactures . . . aided . . . by little steam-engines . . . Carriages, combining lightness and strength, are made here in great number, and exported to all the seaboard of North and South America. There is nobody very poor and nobody very rich." This was not entirely true: carriages accounted for a number of family fortunes in New Haven and, to a lesser extent, in Bridgeport too; although threatened by bankruptcy at the outbreak of the Civil War, the industry switched to Army wagons and prospered right down to the end of the century, bowing finally to the gasoline buggy.

*   *   *

Another venerable and once great industry that disappeared several decades later than carriages was hatmaking. At one time over seventy factories comprised

Sturdevant Woolen Hat Factory, Danbury, 1858

O. Benedict & Company, fur hat manufacturers, Bethel

this industry, fifty in Danbury alone, and most of the rest in Norwalk. The annual production of men's hats reached a peak of 36 million dozens in 1909, the derby then being the favorite headgear. But the advent of the automobile and the increasing fashion of wearing no hat at all caused a decline that culminated in the 1950s, when hatting ceased to be the principal source of jobs in Danbury.

By 1800, not long after the pioneering of Zadoc Benedict and Oliver Burr, hatmaking in Danbury had become the number one occupation. Like the blacksmith or shoemaker, the hatter employed only a few men. Fur had to be imported from England and Germany and processed by hand. Around 1825, E. Moss White invented the first fur-cutting machine for his brother. Most hats were sent in the rough to New York for finishing and trimming, although some were peddled directly. During the 1830s the silk hat, a Chinese innovation, replaced the high muskrat or beaver variety and, in turn, was superseded by the soft felt hat.

A thirty-eight-year-old Danbury farmer named Ezra Mallory did for hatting what Seth Thomas did for Connecticut clockmaking—establish a company so firmly that his heirs were able to perpetuate it for a century and more. A mere apprentice himself, Mallory built his shop close to his farmhouse in 1823. Rabbits and Canadian fur-bearing animals were his chief sources for skins. Two assistants and Mallory cut the fur from pelts with long-handled shears, pulling out the hair with their fingers. Beaver then cost only $4.75 a pound, and Mallory could sell his beaver "plug" hats for $8.00.

The era of mechanization brought labor pains that were to plague the hatting industry through the rest of its history. When Ezra's son introduced Howe's sewing machine to his women employees at the start of the Civil War, they refused to use it until his own sister-in-law shamed them by demonstrating its simplicity. This opened the door for other machines for mixing, blowing, forming, stretching, blocking and felting—each of which met with some resistance. Soon however, it was possible for Mallory to complete a hat in one minute! At the time of incorporation in 1856, E. A. Mallory & Company employed nearly 100 men and women and sold 8,640 dozen hats a year.

* * *

Lambert Hitchcock, the Cheshire-born cabinetmaker, was the first of his trade to apply to furniture-making the repetitive production techniques developed by Whitney and Terry. After serving his apprenticeship in Litchfield he set out in 1818—the year of the new state constitution—for remote northern Connecticut and started a wood-turning shop. At first he made chair parts only, many of which were shipped to Charleston, South Carolina, where they were assembled and sold to well-to-do planters with a taste for luxury. Within a few years he

Hitchcock Chair Company, 1826

found he could do better by producing finished chairs in large quantities at prices ranging up to three dollars wholesale. His simple, sturdy design embodied a curved top, seats of rush, cane or solid wood, and stenciled ornamentation. Beginning in 1826, he employed more than 100 men, women, and children in his new factory in the wilderness. Men, of course, did the woodworking; children rubbed on a priming coat of red paint, which was then covered with black to imitate Duncan Phyfe's rosewood; with brushes and quills women applied the decorations—gold bands, a basket of fruit or the horn of plenty. Across the back was always stenciled "L. Hitchcock, Hitchcocksville, Connecticut, Warranted." Hitchcock was probably the first in the country to manufacture the Boston-style rocking chair.

Only three years after erecting his three-story factory the young manufacturer became insolvent as a "consequence of repeated losses and misfortunes" with liabilities exceeding $20,000. No Whitney or Terry when it came to factory methods, Hitchcock ruefully realized the more chairs he made, the greater his indebtedness grew. His assets included 1,500 chairs on hand, 2,000 more held by agents and a large amount of material that had been sent to the Wethersfield state prison for subcontracting. In a burst of youthful temerity, Hitchcock chose this time to get married, but he also continued in the business as general agent, more successfully than before, by taking to the road and delegating production to his partner. He made frequent and long business trips to New York and Chicago, added chests, tables, beds and other furniture to his line of chairs, and by 1832 had satisfied his creditors.

The Hitchcocksville Company was incorporated in 1841 with the founder as president. At the same time he became involved in politics and served as a state senator. Later, on a spur of the Farmington Canal in Unionville, Hitchcock leased a plant for making cabinet furniture, and in 1848 the Hitchcocksville business was dissolved. Lambert died in 1852, but others continued to make chairs in Hitchcocksville until 1864. Eighty-two years later the business was resurrected and once again flourishes.

*   *   *

On two branches of textiles—carpetmaking and silk—Connecticut early stamped an indelible mark. Like silk, carpeting is one of the world's oldest luxuries, dating back 1,000 or more years before Christ. Also like silk, carpetmaking involved highly skilled and closely guarded processes. The British were the first to convert the art to large-scale production, but Americans, by inventing and perfecting the power carpet loom in 1837, brought the beauty of carpets within reach of middle-class pocketbooks. The output of American carpets in the nineteenth century grew to be the largest of any country. Until after the Civil War, ingrain—with a flat weave and smooth surface—was the dominant type. "It was composed of two or three webs—or layers of material—each of which had a worsted warp and a woolen weft. The warp provided the body; the weft carried the colors." The webs were interwoven to make the design. Other types—Brussels, Wilton, Axminster—were known as pile fabrics. Like the other textile trades, carpetmaking was initially done in the household.

In 1824, three Hartford businessmen decided to take a fling at wool manufacturing. William H. Knight was an experienced mechanic; Nathan Allen a prominent real estate owner; Henry Ellsworth, the son of Chief Justice Oliver Ellsworth, an entrepreneur with interests in banking, insurance and steamboats. For over a decade Allen had been using the power of the Farmington River at Tariffville to manufacture iron, wire and cards. He now purchased additional land for his new project. On April 19, 1825, the *Connecticut Courant* reported: "The corner stone of the 'Tariff Woolen Factory' . . . was laid on the 14th inst. in the

Joseph Toy

Orrin Thompson, 1788–1873

presence of above an hundred spectators. . . ." A four-story stone building soon rose, and the partners obtained a charter and invested $123,000. Although the first product was superfine broadcloth, carpeting was soon added, the raw material coming from discarded wool. In 1826, the *Boston Daily Advertiser* commented:

> We have just seen a piece of Carpeting woven at Tariffville . . . by which it appears that Carpets can be made there of any colours, and to any pattern, durable, cheap, and elegant.

By 1832, the factory turned out an average of 114,000 yards of carpeting worth $120,000 annually, for sale mainly in New York, Philadelphia and Baltimore. The business employed 136 men, women and children. Men earned one dollar a day, women thirty-five cents, children twenty-five cents. Despite such a promising start, the Tariff Manufacturing Company met with financial troubles and in 1840 was absorbed by its competitor in Thompsonville.

The Thompsonville Carpet Manufacturing Company developed into the great Bigelow-Sanford Carpet Company a hundred years later. It was the creation of Orrin Thompson, a native of Suffield, who learned merchandizing in Enfield and eventually became a prosperous New York carpet importer and wholesaler. The rapidly expanding market, however, suffered from increased duties on imported carpets set by the tariff of 1824. In 1828, the unusually dynamic and majestic-looking Thompson persuaded his Scottish supplier to furnish yarn for finishing in the United States. In partnership with his brother Henry and brother-in-law he secured a charter from the legislature, invested $35,500, constructed a dam on Freshwater Brook in Enfield, and set up an integrated production line from raw material to finished product for making carpets and nothing else.

Thompson's Scottish friends further obliged by sending over equipment for spinning woolen yarn and a nucleus of skilled weavers, who received a bonus for staying two years. The first contingent of the illustrious Scottish colony in Thompsonville arrived before the factory or the village was finished. Boarded in taverns used by river boatmen, they helped to complete the dam and worked on the houses known as Scotch Row and on the factory itself. The White Mill, a three-story affair with a cupola and bell and a brook flowing through its basement, looked more like a Congregational church. Ten looms occupied the upper floors, cards and spinning jacks the first two floors. The wool was washed outside and spread on the ground to dry. Around 1832 a "Black Mill" was also erected to provide capacity for three-ply carpeting.

Profits did not roll in as fast as the original investors had anticipated, and Thompson gradually acquired more stock by buying them out. Yet, between 1830 and 1846, the company earned $386,388, giving the stockholders a return of just under 7 percent per year. With the stimulus of the tariff of 1824, carpets from Thompsonville and other mills replaced rag rugs in most public rooms. Thompsonville's production alone amounted to 7,000 yards weekly, three times greater than that of the Tariff Manufacturing Company.

The invention of the power loom by Erastus Bigelow for the country's largest carpetmaker, the Lowell Manufacturing Company, threatened the company's survival and caused Henry Thompson, then agent and treasurer, to withdraw from the business. He felt too old to fight the Lowell interests and opposed the immense outlay that would be required to compete. His brother Orrin solved the dilemma neatly by making a deal with Lowell for 125 to 150 power looms in return for a royalty. Even with this great advantage the Thompsonville mill failed in 1851. Operating losses and bad debts had eaten away the surplus, while outlays for the new power looms were extremely heavy. The preferred creditors took over the ownership, and Thompson himself, then almost sixty years old, carried on as superintendent for another three years, when the Hartford Carpet Company was formed.

The Scottish weavers, dyers and machinists who settled in Thompsonville, among the first wave of labor emigrants from Europe, made up a tightly knit community of proud artisans. They were Presbyterians in religion and Whigs in politics, favoring tariff protection. Paternalism did not appeal to such independent outlanders, who shocked the native farmers by merrymaking on New Year's Day. Although the company provided housing, as in most factory villages, the workers opposed its attempt to increase the board and won. Orrin Thompson, like Humphreys and Collins, showed great interest in his employees' comfort and welfare, but they resented any sign of condescension. When Thompson gave fine broadcloth to the men and silk to the women for Christmas, the spirited workers returned it with the comment, "If ya gae us our proper wages, we'll buy our own suits and dresses." The weekly wage then ranged from four to seven dollars, out of which came two dollars for room and board. Those refusing to board where Thompson assigned them were forthright dismissed.

The workers took full advantage of their skill monopoly, bolstered by a tradition of unionism. They maintained they "had broke factories in the old country and could break this one." Soon battle was joined over wages and working conditions, leading in 1833 to the first strike of workingmen in Connecticut. The men objected to what today would be considered reasonable restrictions, such as those against smoking or reading newspapers on the job and unauthorized absences over fifteen minutes. They also disliked the fines levied for failure to weave a piece (ninety yards) of fine carpeting within twelve days, although premiums were offered for beating the standard. Finally, the introduction of new and fancy fabrics caused the ingrain hand weavers to demand an increase in piece rates.

Management replied by closing the plant and reducing wages. The sixty or so workers reacted by deciding to "stand permanently." Heeding a plea from their leaders to conduct themselves decorously and to avoid the rum barrels, they set up a strike committee, corresponded with fellow weavers in other companies and started paying weekly benefits to members. Arrest of the leaders, bringing in strikebreakers from New York and the threat of eviction from their homes caused the strike to collapse within a month. The men had to return to their jobs at the lower rates. The company failed only to get the strikers convicted of conspiracy. Undaunted, the Scotsmen struck twice more before Bigelow's power loom ended their bargaining strength of hard-to-replace skills. The unskilled men and women who took their jobs did not become sufficiently industrialized to resist management's complete control of their lives until the advent of modern labor organizations a hundred years later. Nationally, unionization during the Jacksonian period shot up in the space of three years from 26,000 to some 300,000, or one-half of the skilled workers in the country. The panic of 1837, however, crushed the labor movement for a long time.

* * *

Cheney Brothers of South Manchester were the first to master the intricate art of silk weaving in the United States. The eight Cheney brothers, descendants of a clockmaking and farming family, after failing as mulberry tree growers, were inspired in 1838 to start a silk mill beside a small brook in an isolated village a few miles east of Hartford. In this rural setting without natural advantages of any kind, their Yankee ingenuity and perseverance created the largest silk complex anywhere that at its peak employed some 5,000 hands and covered thirty-six acres of floor space, surrounded by what was widely hailed as a model manufacturing village.

In Connecticut raw silk production around Mansfield, not far from South Manchester, had increased yearly from the Revolution until 1830, when 3,200 pounds were being produced. Prospects for a national silk industry stirred the interest of Congress, which placed a protective duty of 40 percent on sewing silk and prompted the Secretary of the Treasury to prepare a detailed manual on its growth and manufacture. State legislatures passed laws to help make silk growing pay, as Connecticut had done in the 1780s, and for a decade the outlook enticed many eastern farmers to speculate on growing the mulberry tree.

The Cheney family began experimenting with the *morus multicaulis* species, as tree prices boomed. They rose from $4.00 a hundred to $10, to $30, and for a short while to $100 and more. To meet the demand, the brothers imported thousands of mulberry trees from China and France. They also set up a nursery and cocoonery on leased land in Burlington, New Jersey, and on Charles Cheney's farm in Mt. Pleasant, Ohio. The thriving of over 100,000 trees on their Burlington silk plantation was advertised as "a sight worth seeing". Their monthly publication, *The Silk Grower*, urged others to undertake "one of the most pleasant employments that ever was conducted under the sun. The lame, the halt, the widow and the orphan can perform the labor. Then why not permit them to do so and pay them a part of the $20,000,000 which is annually sent abroad in exchange for silks and to feed foreign mouths and enrich foreign aristocrats?" The road to prosperity seemed quick and easy, but the panic of 1837 did not spare the silk nurserymen any more than it did other enterprises. From his farm Charles wrote his father-in-law: "I had rather be a tenant of the meanest log cabin in Ohio, with health and 'corn-dodgers' enough to eat, than the greatest nabob of a merchant, manufacturer etc. in all New England, to be 'bamboosled' about by the Shylocks and petty Bank Directors that one in extensive business is subjected to in such times as these." Two years later he still grew trees despite meagre sales and abysmal prices. His wife finally admitted failure to her father: "The bubble has burst, and we must go to work again, at the plough, and the dairy, the henhouse, and the pigstye . . ." The brothers lost their entire investment; three of them—Frank, Ralph and Rush—were declared bankrupt.

From the beginning their dream was ill-conceived. The mulberry trees themselves could not survive the northern climate, and a devastating blight killed most of them in 1844. Americans were no match for foreign drudges. Sericulture was essentially a household and manual industry, requiring infinite patience and manifold labor, day and night. The Chinese, then the leading silk producers in the world, received at most three cents for a day's work, almost as much as an American farm girl earned by the hour.

In the meantime the Cheneys had begun using their silk-growing experience to manufacture spool silk for sewing by hand. On January 1, 1838, Frank, Ralph and Ward Cheney, together with their close friend Edwin H. Arnold, formed the Mt. Nebo Silk Mills with a capitalization of $50,000. Ralph was elected president, but Ward actually ran the business, handling all sales and purchases. Near the family homestead on Hop Brook they built a small factory, the water power coming from the bottom of the tailrace of an old grist mill. On March 31 Arnold recorded in his diary: "Raised the silk factory. Had a great many to help." Silk twisting was not unlike a rope walk: the threads were fastened at one end and twisted by turning a wheel thirty feet away. The Cheneys' artist brother, Seth, sent raw silk from France, and with the help of six young farm girls they twisted about ten pounds of sewing silk a week. The

Cheneyville

The Cheney Mills in their heyday

collapse of the mulberry tree venture caused the mill to close for a while, but Seth and John, another artist, came to the rescue financially, and the mill reopened in 1841. It seems an ironic twist of fate that without the early investment of a modest portion of earnings from the artistic endeavors of both Seth and John, the Cheney silk business would probably never have survived infancy.

By the end of 1843, under the name Cheney Brothers, the company had eighteen employees who worked the standard twelve hour day, six days a week. Men earned an average weekly wage of $3.35, women $2.54. The manual operations were tiresome. As the spinning room filled up with farm girls who toiled away skeining the silk, the more literate relieved the monotony by reading to the others. Their favorite material was the *Weekly Courant, Littell's Living Age* and *Uncle Tom's Cabin.* The girls worked or not as they pleased, always going home for holidays. The factory whistle regulated the daily life of the village, blasting at dawn, noon and dusk.

From Edward Valentine in Northampton, Ward learned the mysteries of silk dyeing, which enabled the Cheneys to add colors to their line. In the aftermath of the bankruptcy, long overdue accounts and dunning letters pursued the brothers, but demand for their product steadily increased. A small sales room was opened in New York, which Edwin Arnold managed for many years. Since Italian silk was considered foremost, sewing silk for a long while sold under Italian names; Cheney's best brand went by the name of "Fratelli Chinacci", in black, drab and high colors.

With Frank Cheney's invention of a power spinning machine that combined doubling, twisting and winding, patented in 1847, production took a leap forward in both quality and quantity. Singer's home sewing machine later widened the market for sewing silk, which had to be stronger and more even. In 1855, another major achievement came with Frank's perfection of spinning waste silk. At first, all raw silk had been reeled by hand only from perfect cocoons; now pierced cocoons and waste from unwinding perfect ones, which had been previously discarded, could be used, leading in a few years to ribbons and grosgrain silk dresses at popular prices. A ribbon mill was erected in Hartford on Morgan Street, the company's capitalization increased to $400,000, and a dividend of $50,000 declared. The directors included the five active brothers—Charles, Frank, Ralph, Rush, and Ward—plus E. H. Arnold and Frank W. Cheney, the son of Charles. In 1857 removal of the raw silk duty gave a powerful impetus to domestic silk manufacturing. Over half of the silk bought was then made in American mills. Of more than 500 mills, Cheney Brothers had already become the largest and probably the most profitable.

While his brothers were for the most part sparing of speech and reserved in manner, Ward Cheney had a warm, generous character and immense personal magnetism. His executive ability quickly elevated him to the head of the business, even before he assumed the title of president, which he held until his death in 1876. Ward's buoyant nature had much to do with the family sticking together after the mulberry tree fiasco and carrying on the silk mill. Naturally friendly to all, he established the good relations with his associates that later gave Cheney a national reputation for its humane treatment of employees and made South Manchester a manufacturing utopia. On wintry days when the snow lay deep, he drove his sleigh to collect the mill girls from their homes.

A curious sidelight to Ward's career was his devotion to spiritualism, which caught the East's fancy in the middle of the nineteenth century. During this mass psychic disturbance, mediums were materializing luminous faces and hands, producing music, voices, lights and icy currents of air, as well as speaking in strange tongues. Not far away from South Manchester, in Norwich, a seventeen-year-old Scottish boy named Daniel D. Home (pronounced Hume) not only had visions but apparently caused the furniture in his aunt's house to shift about surreptitiously. Soon he was lionized all over New England by well-to-do merchants, doctors, editors and clergymen. He performed before Thackeray at the home of the historian George Bancroft and amazed Horace Greeley.

Ward Cheney frequently invited Home to visit the Homestead, where on August 8, 1852, before Ward and Rush Cheney and Franklin L. Burr, editor of the Hartford *Times*, he gave the first of his rare feats of levitation. According to Burr, the gentlemen had adjourned to a darkened room to see whether spirit lights might not shine when unexpectedly Home began his ascent. Burr said: "I had hold of his hand at the time, and I felt his feet—they were lifted a foot from the floor!" Another night he held a seance around the mahogany dining room table; Cousin Anne Cheney, then a little girl, was in the midst of eating her pudding when the table started to rise as much as two feet. She cried out, not from fear, but for her suddenly removed dessert, at which point Home made the table drop and spin around until her plate appeared in front of her.

Two years later, in 1855, Home journeyed to England, where Elizabeth Barrett Browning wrote in her *Letters*: "The American medium Hume is turning the world upside down in London with this spiritual influx." The scientist Thomas Huxley is reputed to have witnessed Home float out of a second-story window to an adjoining one. Now famous, he delighted the courts of Europe, finally marrying a Russian princess and spending most of his life in Moscow. The very day Ward died he penned, from a distance of 5,000 miles, condolences to the Cheney president's daughter-in-law.

The dominating personality in Cheney Brothers for a generation after Uncle Ward's death was Frank Woodbridge Cheney, the son of Charles. "When he sat at the head of the table on Thanksgiving Day," wrote his granddaughter, "surrounded by twelve children, their families, cousins and aunts, over seventy by count, he was the focus of all attention. The newest born would be seated on his right, briefly, and held by him a little while." His penetrating blue eyes—a Cheney characteristic, snow white beard and white hair, gold-headed cane and watch chain, and his immaculate appearance—all enhanced the image of a patriarch feared and respected by family and employees alike. Frank W. became a director of the mills in 1854 at the age of twenty-two. He had attended Brown University but was expelled for breaking Puritan tradition by going to a performance in Providence by Jenny Lind, the Swedish Nightingale, on the Sabbath.

Confronted by the removal of the raw silk duty and the high cost of European materials, the elder Cheneys decided the time had come to establish trade relations in the Far East. China was then the leading producer of silk. Commodore Matthew Perry's bold entry into Japan had recently resulted in a treaty allowing Americans to trade there. As the most promising member of the younger generation, Frank was dispatched to China by way of France, armed with £50,000 of credit—equal to five years of Cheney profits. He arrived in Shanghai in the spring of 1859, as the Treaty of Tientsin opened doors deep into China and the aging Manchus were challenged by the Taiping rebels. While he was traveling upriver to visit the silk districts, pirates boarded his Chinese junk. The malevolent leader was first over the side, musket in one hand and a lighted punk to ignite the powder in the other. Frank stepped forward, took the punk from the pirate's hand and lit his cigar, then handed another to his adversary and lit that. His courage won the day; the two men sat down as friends and communicated as best they could.

Frank quickly learned that picking out a year's supply of silk was no simple matter. Under the "hong" system foreigners had to live in compounds called "factories" and deal exclusively with a limited number of merchants. He wrote home:

> You cannot run around but must sit still and have everything come to you. Such is the custom in China and in China the custom is the law . . . You are often obliged to buy a chop of silk as you do a bushel of apples at home and you take the speckled ones as well as the fair . . . The Chinese are very hard working and patiently industrious race, but a horribly dirty one . . . The stinks are on all sides . . . They are a great set of rascals, and are certain to keep as far from the truth as possible.

He also found the currency perplexing and transportation painfully slow; to reach the United States by sea took four months. Silk exports to America then amounted to but a small fraction (2500 bales) of the 90,000 bales annually received by England. Soon the difficulty of becoming accepted by the Chinese made him realize the necessity of working through established agents, especially the house of Augustine Heard & Company, whose friendship he won and with whom he resided.

A few months later he sailed to Japan. His impressions of that newly discovered civilization were perhaps the first to be recorded by an American businessman in letters to his father and uncle. Entranced by the natural beauty of Nagasaki and its hills, compared to the oppressive flatness of coastal China, he preferred the people as well, finding them clean, free from vice and more vivacious. The quality of their silk almost equalled the very best. He was well aware, too, of the "fatal impact" of opening up Japan, a feeling shared by the insular government which did its best to restrict trade:

> I fear they will reap more evil than good from their intercourse with the so-called civilized world for some years to come . . . All trade is, as yet, carried on in a very primitive way with the Japanese. I have had to do all the buying, inspecting, paying (in Mexican dollars), packing and shipping myself . . . Silk in Japan, as it is everywhere else, is an article of luxury and consumed by the hereditary princes of the land in whom the ruling powers are vested. They have, of course, already begun to feel the effect of foreign demand for silks and the great advance in prices—which touches them in a tender spot—their pocketbooks. Consequently, the government is now trying to prevent export . . . But trade goes on in spite of government edicts and two-sworded officials . . .

> The Japanese are not such a hard, shrewd and practical money-making people as the Chinese, but are infinitely more pleasant people to live with. They live and act just like children and there is a great deal of charm about their free, easy, natural and good natured manner. I am afraid this will not last long. Their intercourse with foreigners will not improve their morals or manners. I have great faith in the Japanese and hope the Anglo Saxons will not crush them as they have all the other weaker races.

Uncle Ward, delighted with the samples of Japanese silk, among the first to reach America, urged him to buy all he could lay his hands on.

Meanwhile, in South Manchester, the brothers were ecstatic about their success in spinning waste silk, enlarging both the mill there and the new one in Hartford, and replacing sewings and twist with weaved goods and pongees. Frank's father, apparently overburdened by his responsibility for the expansion, yearned for his son's return and expressed concern over the business's rapid growth: "They have at Manchester good mechanical and executive ability. They sometimes with their never ending talk get into soup and do not sail along quite as smoothly as they might."

On his third trip to Japan, early in 1860, Frank bought sixty-one peculs of raw silk at a cost of $24,765 and advised his father that he intended to stay in the East another season so as to take advantage of the lower Japanese prices. But already he was disillusioned over their honesty: "The Japanese are far ahead of the Chinese in the art of cheating and false packing of their silks." During the year silk prices there rose substantially, most of the increase, instead of benefiting the Nippon merchants, going for government squeeze. His father echoed Ward's pleasure over the shipments home:

> Nothing can do better than the Japan silks—with such silk we can run all the other manufacturers off the mark . . . Our exports with China and Japan will amount to a formidable sum in the next ten years. We shall have a railway to the Pacific in less than fifteen years (actually, 1869) and a steamer from San Francisco to China and Japan in less time . . .

In October, Frank, commenting on the march of Allied forces toward Peking, told Uncle Ward: "This is probably the last dying struggle of the old dynasty . . . The renovation of China, like its decay, must be the work of centuries." He felt the Chinese revolution would disrupt the silk districts for a long time. As the clouds of civil war also gathered in the States, he wrote: "I shall say goodbye to China . . . come back to Hartford and become a respectable member of society." Keeping his promise, he returned home, joined the Army and married Mary Bushnell, daughter of Horace Bushnell, the famous liberal preacher.

As complete and unique as the Cheney mills was the village life surrounding them. New England mill towns, especially after 1860, were generally noted for grimy rented hovels, unschooled child labor, worn out and poorly paid workers, and almost intolerable working conditions. Industrial feudalism, many contended. But a different kind of seigniory existed in Cheneyville:

> At South Manchester . . . there is . . . the most attractive mill village in the country . . . The Cheneys are not only the employers of their operatives, but their neighbors as well . . . A visitor . . . would find it difficult to believe that large red brick factories exist amid so much natural and artificial beauty. The village is scattered through a large park, . . . The grass is beautifully kept. From one end to the other of the park there is not a fence . . .
>
> The Cheneys do not like to have South Manchester spoken of as a model village, or that they are trying an experiment . . . The ruling motive of the men who have built up South Manchester was that the village should be an agreeable and pleasant home for themselves and their operatives . . . There is a free library and reading room established in the former house of one of the elder Cheneys . . . Churches have grown up . . . nearly all the sects are represented . . . When there is a

family gathering more than a hundred Cheneys assemble, and their fixed feasts are held in the (Cheney) hall.

As the business grew, and houses for the operatives were needed, the mill owners put them up in the park . . . They are comfortable cottages, much better than the ordinary factory tenement . . . Most of the operatives in a silk mill are girls. They are obliged to board . . . Two large boarding houses have been maintained . . .

The secret of the success of South Manchester as a village lies in the fact that the proprietors . . . constitute a very large family directly interested in the mills . . . who take an active and intelligent interest in public affairs . . . Not a single strike is reported as occurring . . .

For a company house, Cheney employees paid about eight dollars in monthly rent. The mills were well ventilated, with plenty of windows. Besides the library, the Cheneys supported a school; operated a 700-acre farm, store and grist mill; constructed several large reservoirs to provide drinking water; financed a fire department; formed a light, power and horse tramway company; provided sewers and garbage collection; and maintained all the streets in the mill district. Cheney Hall offered interdenominational religious services, theatrical and musical entertainment, lectures and exhibitions. What had started as a business enterprise unintentionally turned out to be a successful social experiment, simply because the Cheneys wanted to work where they lived and liked keeping a tidy backyard and being good neighbors.

It took the Cheney family eight decades to build the greatest silk company in the United States, to make Cheney silks the best in the world, to establish a harmony with labor rarely equalled anywhere, and to develop around their mills a community with a soul. Yet their empire disintegrated in less than a decade. The dynasty, which for 117 years had shown unusual cohesion, shrewd management, artistic integrity and social consciousness, died from inability to adapt old habits and traditions to the realities of the style-conscious mass market that America became.

* * *

Nine years after the successful start of the Cheney Brothers in silk, a young gunmaker, Samuel Colt, returned to his native city to achieve his life's ambition of having his own arms factory. After having been a failure at school and in business for two decades, although not as an inventor and pitchman, he was to make his mark in Hartford, in the world and on posterity almost at once.

To many in Hartford, Sam's brash nature and new-fangled ideas made him seem an outsider—a wild frontiersman rather than a sensible Yankee. Yet his grandfather on his mother's side, John Caldwell, had founded the first bank in

Hartford; his father was a merchant speculator who made and lost a fortune in the West Indies trade. Widowed when Sam was only seven, the year the boy took apart his first pistol, Christopher Colt had to place his children in foster homes for a while. At ten, Sam went to work in his father's silk mill in Ware, Massachusetts, and later spent a short time at a private school in Amherst. At Ware he learned something about chemistry and electricity and fashioned a crude underwater mine by a combination of gunpowder, electricity and wire covered with tarred rope. Here he also became acquainted with a young machinist named Elisha K. Root, the same man who would someday run Colt's great armory.

In 1830 he persuaded his father to let him go to sea. A Boston friend arranged for him to work his passage on the brig *Corvo*, bound for London and Calcutta with a cargo of Yankee cottons and missionaries. During this, his sixteenth year, he conceived, by observing the action of the ship's wheel, or possibly the windlass, a practical way for making a multishot pistol. Probably out of a discarded tackle block he whittled the rotating cylinder designed to hold six balls and their charges. His inspiration was to enable the pawl attached to the hammer of a percussion gun to move as the gun was cocked, thus turning the cylinder mechanically. Colt thus became the inventor of what was eventually the first successful revolver.

Returning to Boston with the model, he managed, with his father's aid, to have two prototypes fabricated by a Hartford gunsmith, Anson Chase. One failed to fire; the other exploded. Now out of funds, Sam summoned all his ingenuity to make his living and save enough to continue the development of his revolver, which he alone was certain would make his fortune. At Ware his exposure to chemistry in the silk mill's dyeing and bleaching department had introduced him to nitrous oxide or laughing gas, discovered by Sir Humphrey Davy in England about the turn of the century and later used for the first time as an anaesthetic by a Hartford dentist, Dr. Horace Wells. Soon Sam set himself up as the "celebrated Dr. Coult of New York, London and Calcutta" and for three years toured Canada and the United States, billed as "a practical chemist" giving demonstrations, for which he charged twenty-five cents per head. The gas intoxicated its inhalants for about three minutes, under which condition they performed ludicrous feats to the delight of the audience.

In the meantime Colt had hired John Pearson of Baltimore to make improved models of his revolver, but he was still at wit's end to keep the constantly grumbling Pearson going as well as himself. Borrowing $1,000 from his father and living on a shilling a day, he journeyed to England and France and obtained patents from both countries. In 1836, aided by the U.S. Commissioner of Patents, a Hartford citizen named Henry Ellsworth, he received U.S. Patent

No. 138, on the strength of which he persuaded his conservative cousin Dudley Selden and several other New York men of means to incorporate the Patent Arms Manufacturing Company of Paterson, New Jersey, and to invest about $200,000. For his part Sam got an option to purchase a third of the shares issued (although he was never able to pay for one of them), an annual salary of $1,000 and a sizable expense account, of which he took full advantage to promote his five-shot revolver in Washington military and Congressional circles. At the time the Army Ordnance Department, facing boldly backward as always, was smugly satisfied with its single-shot breechloading musket and flintlock pistol. A West Point gun competition rejected Colt's percussion-type arm as too complicated, and Cousin Dudley grew impatient with Sam's debts, lavish dinner parties and lack of sales. In 1842 the Paterson company closed its doors for good.

Colt ended up in debt and in controversy with his employers, whom he suspected of fiscal skulduggery. Disgusted with bureaucrats, he determined to be his own boss in the future. To a relative he confided in his colorful, half-educated way:

> I have never forgot a saying almost the first that I remember in life at least is among the most impressive 'It is better to be at the head of a louse than at the tail of a lyon' . . . if I cant be first I wont be second in anything.

Even before the demise of the Paterson company, Colt had been working on two other inventions with a little more success. On a visit to Florida he had seen what moisture could do to gunpowder and he developed a waterproof cartridge out of tinfoil. After its endorsement by General Winfield Scott, Congress, in 1845, amended its annual appropriation of $200,000 for arming the militia of the states to expend a quarter of this amount on Colt's waterproof ammunition. This achievement, in turn, led him to resurrect and refine his Ware experiments with underwater batteries. What better and cheaper way to protect the Eastern seaboard from enemy men-of-war! The Navy granted him $6,000 for a test. Using copper wire insulated with layers of waxed and tarred twine, he made four successful demonstrations, one of which blew up a sixty-ton schooner on the Potomac before a host of Congressmen and government officials. But neither the military nor Congress took to his idea, and his Submarine Battery Company never surfaced.

In 1846, the same month war was declared on Mexico, the New York and Offing Magnetic Telegraph Association was incorporated by Colt and a new set of business associates, with the rights to construct a telegraph line from New York to Long Island and New Jersey. But again the operation was mismanaged, partly due to Colt's negligence, and at thirty-two he once more found himself as "poor as a churchmouse."

Although Colt was not destined to fight in the Mexican War, his guns did. The five-shot Paterson pistols, having won acceptance by the troops against the Seminoles in Florida, gained further renown in the hands of the Texas Rangers in the early 1840s. At the end of 1846, without money or machines but still possessed of his patent rights, the inventor approached the Ranger Captain Samuel H. Walker about buying his "improved" arms for his men, who had been mustered into the U. S. Army. A veteran Indian fighter, young Walker wrote Colt that he thought with improvements his pistols could be made the most efficient weapons in the world. Yet he had to admit that nine out of ten government officials in Washington still did not know what a Colt pistol was. Soon the slender and modest Walker and the ebullient inventor met and became fast friends. That winter General Zachary Taylor, now stationed in Texas, wanted 1,000 Colts within three months, but Colt lacked even a model with which to start manufacturing again. Together, he and Walker toured the New York gun shops without locating a single Paterson-type pistol for sale. This did not unduly distress either of them, however, because the captain wanted a simpler yet heavier gun (.44 calibre) that would fire six shots, and Colt designed, mostly from memory, the so-called Walker gun.

Armed with a government order for $25,000, Sam persuaded the Connecticut contractor for U. S. muskets, Eli Whitney, Jr., to make the thousand new revolvers, which were finished six months later. His appetite whetted, Colt obtained a second order for another thousand Walker-type guns, but this time he was determined to depend upon nobody for the finished product. He borrowed about $5,000 from his banker uncle Elisha Colt and other Hartford businessmen; hired several scores of hands; and using machinery, tools and surplus parts from Whitneyville, which were his by contract, started his own factory in Hartford during the summer of 1847. He promised to supply 5,000 arms a year. To a friend he wrote:

> I am working on my own hook and have sole control and management of my business and intend to keep it as long as I live without being subject to the whims of a pack of dam fools and knaves styling themselves a board of directors . . . my arms sustain a high reputation among men of brains in Mexico and . . . now is the time to make money out of them.

Alert to the new methods being used at Whitneyville and other armories, Colt quickly adapted the Whitney system of interchangeable parts to the mass production of revolvers. He perfected it to the point where he could claim that 80 percent of his gunmaking was done by machine alone. Vital to his success was his ability to hire the ablest machinists and managers of the day, especially Elisha K. Root, the mechanical genius whom he had first met in Ware. Offering Root

Colt's Patent Fire Arms Manufactory, Hartford

Inside of Colt Armory

the then-unheard-of salary of $5,000 a year, Colt lured him away from the famous Collinsville machete company. As head superintendent, Root conceived of many of Colt's marvelous belt-driven machines—some of which run to this day—for turning gun stocks, boring and rifling the barrels and making cartridges. In a few years he would design and construct the incomparable Colt Armory and install its equipment. Under Root's quiet but firm leadership, one that despised sham and sought perfection, Colt's became a training center for a succession of gifted mechanics, who went on to apply its manufacturing techniques or to head companies of their own—men like Francis A. Pratt, Amos Whitney, Charles E. Billings and Christopher M. Spencer.

While Root managed the factory, Colt as president filled the role of salesman extraordinary. Far more than his competitors he appreciated the necessity of creating demand through aggressive promotion. From 1849 to his death he devoted much time to traveling abroad, wangling introductions to heads of states, and making them gifts of beautifully engraved weapons. At home, military officers and others acted as his paid agents in the South and West and as his lobbyists in Congress. In perhaps the first use of art in advertising, he engaged the noted Indian painter, George Catlin, to do a series of canvases depicting the Colt gun in a variety of exotic settings. Until the approach of the Civil War, however, government sales were scanty compared to the thousands of revolvers shipped to California during the Gold Rush and to such mutual enemies as the Sultan of Turkey and the Czar of Russia. Colt's success in obtaining a seven-year extension of his basic patent and in crushing attempts at infringement now placed him in a triumphal position that made him a millionaire in less than a decade. Finally, as a loyal Democrat he won his long-sought colonelcy, becoming an aide-de-camp to his good friend, the newly elected Governor Thomas Seymour, a former Congressman and hero of the Mexican War.

Leaving Root to run the Hartford works, Sam, still a high-living bachelor at thirty-five, his six-foot frame heavier and with wrinkles around his light hazel eyes, set out to captivate England and the Continent. In May of 1851 at London's Crystal Palace Exposition he exhibited 500 guns and served free brandy. Fascinated by the obviously superior quality of these machine-made arms, the London Institute of Civil Engineers invited Colonel Colt to read a paper. He called it "Rotating Chambered-Breech Firearms" and dwelt, with typical Colt embellishment, on the success of his repeating pistols in exterminating the Indians and Mexicans. In 1853, Colt became the first American to open a branch plant abroad, choosing a location along the Thames for supplying the English government with what he termed "the best peesmakers" in the world. So backward did he find England's mechanical competence that he was forced to

The artist Catlin shooting buffalos with Colt's revolving pistol

send over both journeymen and machines. He spoke of giving a Thanksgiving party for his Yankees. But even the indefatigable Colt was unable to convince the English that machines were superior to hand labor and would not cause unemployment. Harassed by training problems and left without a market at the end of the Crimean War, for which he provided 200,000 pistols, he sold his London factory in 1857.

Back home, with employment and production continuing to soar, Colt rushed to supply state militia in both the South and North and to meet foreign demand. Forced to seek larger quarters, he dreamed of building the largest private armory anywhere. His attention turned to the 200 acres of lowlands below Hartford, where the Dutch had first set foot, and he conceived of a grand scheme for reclaiming them by building a dike nearly two miles long to contain spring floods. In two years, his dike, with French oziers planted on top to prevent erosion, was finished at a cost of $125,000. The Hartford Common Council fought him every step of the way, and many local citizens scoffed at the venture. Colt's dike survived the record flood of 1854, which reached almost twenty-nine feet, and protected his South Meadows until New England's worst deluge in all history occurred in 1936.

Colonel Samuel Colt

Behind the dike soon rose the great Armory, built with brownstone from nearby Portland, topped by a blue onion-shaped dome and gold ball, with a stallion holding a broken spear in his mouth. A giant 250-horsepower steam engine, its flywheel thirty-feet in diameter, drove the 400 machines by means of a labyrinth of shafts and belts. By 1857, Colt was turning out 250 finished guns each day. The buildings, measuring 500 feet long by 60 feet wide, were steam heated and gas lighted. Around them he constructed fifty multiple dwellings in rows for his workmen and their families, a pattern of streets, a reservoir, and, half a mile to the south, a plant for making his waterproof metallic-foil cartridges. Colt paid the best wages, which then averaged $41.64 monthly, but insisted on maximum effort. By practicing "inside contracting" he kept his own employment to less than a fourth of the total number who earned their living from the Armory. His contractors, thirty-one in all, each assumed complete responsibility for a particular operation or department, hiring their own men and receiving from Colt materials and tools.

The willow trees grew so well atop the dike that Colt set up a small factory to manufacture willow furniture, which became popular in Cuba and South America because of its lightness and coolness. Unable to find trained willow workers, he imported a village of some forty Germans. Anxious to provide them with familiar surroundings, he erected a row of two-family brick houses with outside staircases modeled after their own Swiss-style homes in Potsdam, as well as a beer and coffee garden. Then, both for his own pleasure and theirs, Colt formed the Armory Band, giving its members blue uniforms with caps bearing the Colt insignia and instruments engraved with a revolver. His final contribution to employee welfare was Charter Oak Hall, named after the great Charter Oak tree that fell the same year as the Hall was dedicated. Part of the Armory and seating 1,000 people, it served as a meeting place where the workmen could read, hear lectures or concerts and hold fairs or dances.

Colonel Colt was now at the peak of his career. He lacked only a wife and home, and these he acquired with his usual dispatch and pomp. He chose as his bride the gracious and gentle Elizabeth Jarvis, twelve years his junior. After a wedding trip to Europe he built "Armsmear," one of the half-dozen most elaborate residences in America in its day. When Armsmear was finished, Colt's investment in the South Meadows approximated two million dollars. This he accomplished without borrowing from the bankers he roundly detested, truly a gigantic redevelopment project for that era, the importance of which was lost upon the city fathers. Although Hartford finally gave him some tax relief for his improvements, its only contribution was three street lamps. So exasperated did the Colonel become over such niggardly, hostile treatment, which was undoubtedly aggravated by his own brashness, that he made a major change in his will, depriving Connecticut of what would surely have been its greatest engineering school, rivaling M.I.T. or Rensselaer.

By the end of 1858, as North and South raced toward cataclysm, Colt was making enormous profits by filling the demand of both sides for what he sardonically called "my latest work on 'Moral Reform.' " The Armory's tax-free earnings averaged $237,000 annually until the outbreak of the Civil War, when they soared to over a million. Trying again for a patent extension, he offered a bribe of $50,000 to be divided equally between the Republican and Democratic legislators in Congress if his bill became law. It didn't. Although he abhorred the prospect of war, yet he had enough foresight to prepare his Armory for a five-year conflict and the arming of a million men; the prevailing sentiment in Hartford was that war could not last two months.

Now his heavy business responsibilities—the Armory's million dollar expansion program, government contracts, the willowware factory, an ill-fated silver mining venture in Arizona, western land speculation—all were taking their toll of even Colt's seemingly inexhaustible energies. He drove himself as if he knew his days were numbered. His father-in-law drew a picture of a human dynamo who loved work most of all but could still find time to arrange a sleigh ride and party at Armsmear for the children of his employees. Smoking Cuban cigars, Colt ruled his domain from a rolltop desk at the Armory, often writing his own letters in his left-handed scrawl, poorly spelled but forcibly expressed, and usually signing them "Yours in Haste" or "Your obt. Servt." He died in January, 1862, at the age of forty-seven.

What kind of man was Sam Colt? To his friend and neighbor, I. W. Stuart, "Colt has stamped his character . . . upon his age—has concentrated upon a masterly invention of his own, the eye of the world, and is in himself a living epoch." To Senator James of Rhode Island: "Had not Col. Colt been a man of the most fertile genius, of indomitable energy and perseverance . . . he would long since have been borne down by the hostile forces arrayed against him . . ." He certainly did not square with Bernard Shaw's latter day archetype of the munitions maker in "Major Barbara." Money was not his only religion; as an inventor, promoter and capitalist, he craved success, power and—above all—the distinction of being first and best. Although equally unashamed of profiteering, equally direct and simple in his motivations, he seldom gave a thought to the morality of dealing in death and destruction. But neither did the church nor the social mores of his time. Gunmaking could be no sin to Connecticut Yankees who had made their state an arsenal for the nation since colonial days. What did bother the diluted Puritan conscience of the mid-nineteenth century was the personal life of a Hartford aristocrat, born to Congregationalism, who flouted its tenets by his bizarre career, his love for high living, his overbearing pride and

Armsmear

Aerial view of Colt Armory today. Note former homes of Colt workers in left foreground

flamboyance. Colt had more patriotism than the typical munitions merchant, but even this quality was subordinate to his desire to maintain a status quo between the North and South, his contempt for bureaucratic bumbling and pettiness, his willingness to use bribery to gain his ends.

The first manufacturing tycoon, Colt saw the development of several dominant themes during his career besides the overriding one of slavery; the decline of Puritanism into the veneered Victorian morality; the rise of manufacturing in New England to preeminence; the opening of the frontier, with the help of the Colt revolver; the triumph of the individual entrepreneur, creating the Horatio Alger legend. In this milieu Colt was the foremost of the Yankee peddlers—a practical dreamer, head in the clouds, feet on the ground. The success of his many inventions was due less to their intrinsic merits, which were great enough, than to his showmanship in telling the world about them. He achieved his goal despite constant adversity for nearly three-quarters of his short life. Proud, stubborn, farsighted, he stood above and apart from the crowd, impatient with the old ways, "paddling his own canoe", as he said, and preferring "to be at the head of a louse than at the tail of a lyon." He was both.

*   *   *

There is no single prescription for company survival, but many of the essential ingredients can be isolated. First of all, companies—like any institution—are subject to the universal law of continuity and change. In order to continue, they must adapt to change as circumstances demand. Second, the chances of survival are much greater if the product made is proprietary in nature—one that can be patented initially, improved and patented again, and over many decades so nurtured through managerial ingenuity that competitors find it difficult to challenge. Third—and most important—is the succession of alert, aggressive managers.

One Connecticut company holds title to three unique records of longevity. Dexter Corporation in Windsor Locks, founded in 1767, is not only the oldest manufacturer in continuous existence in the state but also the oldest to have the same family ownership, now in the seventh generation. Moreover, Dexter has a similar standing among papermakers of the United States.

The first Dexters ran the village grist, saw and fulling mills. Charles H. Dexter, of the third generation, in the basement of his father's gristmill, experimented with making wrapping paper from Manila rope, perhaps using old saltpeter bags from Hazardville's powder mills. In 1847, he formed a partnership with his brother-in-law, Edwin Douglas, the engineer who had come to Windsor Locks to superintend the construction of the Enfield Canal. Later Dexter headed the Connecticut River Company which provided waterpower for the town and

Statue of young Colt whittling the model cylinder for his first revolver

The Colt dome, a Hartford landmark

looked after navigation above Hartford. On May 1, 1854, when the great flood of that year had raised the level of the river to such a height that communication with Hartford by land had been cut off, Dexter suddenly remembered he had a note maturing at a Hartford bank. In true Yankee fashion, combining resourcefulness with responsibility, he chartered a small steamboat to take him downriver. Since the bridge at Hartford was submerged, the captain piloted his vessel around the eastern approach, back across the river, right up State Street, and moored alongside Bull's drugstore on Front Street. His passenger then disembarked and paid his note on time.

From Charles Dexter's 200 pounds a day of manila wrappers the company grew steadily into a specialty producer of absorbent and filter papers, including almost all of the porous tea bag papers in the world.

Another Connecticut company that has met the test of time is Ensign-Bickford of Simsbury, the first manufacturer in America of safety fuse, basically a textile operation. In terms of both policy and performance, Ensign-Bickford has few peers in America among medium-size firms. Avoiding the temptation of smugness which has beset others that have enjoyed a similar proprietary position, this family-owned enterprise has survived through five generations. In keeping with Connecticut tradition Ensign-Bickford has made significant contributions to American military and space efforts for well over a century.

Despite its essential role during wartime, Ensign-Bickford, unlike most enterprises, can lay claim to having been founded out of a concern for humanity. When William Bickford, a Cornish leather merchant, invented his "Miner's Safety Fuse" in 1831, he did so not for any material gain but in the spirit of saving lives. The many tragedies he witnessed in the ancient English tin mines fortified his resolve to find some way to make the lives of miners longer and safer. In those days blasting was done with gunpowder ignited by a crude fuse made of wheat straws or goose quills, telescoped together and filled with powder. One day he visited a ropemaker in a nearby fishing village and walked with him along his rope walk. As he observed the twisting process, the idea struck him that if a funnel filled with powder could be arranged to pour a stream into the center of the twisted strands, and if the rope could be securely fastened and waterproofed, he would have a slow-burning fuse. During an early experiment many seconds passed, and still no flame appeared. Had it gone out? Could this simple piece of twisted yarn cause gunpowder to burn so slowly?

It worked. "Sir, 'tis a very tortoise!" his assistant exclaimed. Thus was born the partnership of Bickford, Smith and Davey, and within a few years a spectacular decrease in blasting accidents resulted from Bickford's invention of safety fuse.

Shortly thereafter Richard Bacon, then superintendent of the Phoenix Mining Company in East Granby, where copper had been discovered as early as 1705, read about Bickford's patent. He journeyed to England in 1836 and talked himself into being appointed exclusive agent for the sale of safety fuse in America. Domestic manufacturing was also part of the deal. First at the mine and soon in a small shop next door to his house in Simsbury, Bacon began twisting safety fuse by hand with the part-time help of local farmers. Flax, hemp or cotton was spun around the black powder, then covered with a strong twine wound at right angles—the countering operation—and finally immersed in heated tar for waterproofing. In the eyes of his English partners, Bacon ran a sloppy operation and failed to keep good accounts, causing them to dispatch an affable, ruddy-cheeked young Cornishman to set things right.

A mineworker, Joseph Toy, had educated himself to become a licensed Wesleyite preacher and teacher. In a few years, as Bacon's health began to fail, Toy virtually took charge. The clash of their personalities caused an irreconcilable rift. Self-confident, optimistic, extremely ambitious, Toy worked night and day. Bacon, on the other hand, viewed the Cornishman as a foreign upstart. A typical shrewd, stiff-necked and uncompromising Yankee, ill-tempered from his attacks of rheumatism, Bacon allowed his farm and mining interests to interfere with fuse making. Joseph Toy wrote his English partners: "Mr. Bacon never half worked the business." This mutual distrust boiled over in 1851 after a fire destroyed the factory for the second time. Completely discouraged, Bacon wanted to terminate the partnership, but the indomitable Toy—who had built up a bigger demand for fuse than he could produce before the fire—urged him to rebuild. Bacon procrastinated. Finally, Toy, exercising a secret power of attorney from his English principals, formed another company and for $1,000 purchased a new site on Hop Brook in Simsbury village. Backed by his English partners and retaining his customers, he was able in just three months to ship fuse from his new plant to Boston. In the meantime, Bacon held onto the old factory and vainly tried to carry on by himself. Later he agreed never to engage again in making safety fuse, in guarantee of which he posted a bond of $5,000 payable to Joseph Toy, who purchased his real estate and machinery.

The next year Toy sold 11 million feet of fuse. He introduced new varieties and continuous process machinery. Employment averaged under thirty, top pay was fifty cents daily, the usual long work day broken only by an hour for lunch and in summertime by an afternoon swim in the river. In the two-room brick office Toy kept his desk in front of a Franklin stove. Fuse spinning and countering were done by women, the varnishing by men. The preparation of the waterproof mixture was colorful: tar was melted and boiled in huge kettles heated by wood fires, the kettlemen judging its proper consistency by chewing it. Youngsters were taught to coil the fuse and pack it into secondhand barrels from the local grocery store. Until well into the nineteenth century, before railroads spanned the continent and long before our modern highway system had been conceived, horses and steamboats were the only means of transportation. Twice a week fuse shipments were driven twelve miles over the hills to the port of Hartford, where they were loaded aboard side-wheeled packets bound for New York or Boston. In those days Joseph Toy used to exclaim: "If we catch the boat at Hartford, we'll get the order!"

In a company that necessarily relies on powder as a principal ingredient of its products, the danger of fire or explosion is an ever-present spectre that haunts the working lives of those involved. The first quarter century saw three catastrophes, all fires, the last of which served to stamp the rule of safety first indelibly upon the minds of management. This took place early on the morning of December 21, 1859, in the fuse spinning room, dangerously located on the second floor of Toy's wooden building. The next day the *Hartford Courant* carried this headline: "Terrible Calamity at Simsbury! Eight Persons Burned to Death!" The persons were all females, including two fourteen-year-old girls; four more suffered severe injuries. The hands of Toy's son were badly burned, but he escaped possible death by putting his head through a window and jumping out into the flume. The factory was totally consumed. The cause could have been the coal stove that stood at one end of the fuse room, or sparks from the machine shop and blacksmith forge below. Commented the *Courant*: "There was no insurance, of course, as Insurance Companies refuse risks on such property." Then it was that Toy realized the sine qua non of cleanliness and safety; thereafter he built only with brick and stone, decentralizing all operations into small, one-story structures well separated from one another.

By the nature of its products and operations, Ensign-Bickford bears no comparison to the usual factory. It has maintained a tranquil, campuslike atmosphere right in the middle of an unspoiled village. Scores of neat, white homes, very different from the usual mill-town row dwellings, line both sides of the highway into Simsbury, most of them until recently company owned. Further on stands a group of one-story red sandstone buildings, between which courses the usually quiet Hop Brook. The oldest existing structure was built about 1860, nine years after the plant located on its present site. As the village's largest property owner, with most employees and officials living within a mile of the factory, Ensign-Bickford had to be a good employer, the leader in town affairs, the conscience of the community. From the time Joseph Toy settled in Simsbury an intimate relationship prevailed between worker and owner. Soon after his arrival, his sermons were in great demand by the local Methodists; it is said he preached in the schoolhouse at Weatogue the evening of his first day in Simsbury. All of his children worked in the shop, usually during summer vacations, a custom continued in each succeeding generation of the family. With his religious and pedagogic background, Toy created a fatherly, helpful image that was continued and polished by his successors.

For its first hundred years Ensign-Bickford functioned as an informal partnership like so many early companies. Almost daily the executives met in the old three-room office with its open door. Each officer participated freely and fully in all phases of the business, without any recognized division of responsibilities. But control remained in the hands of the same founding families, which achieved a consistent evolution of sound principles and policies and provided a superior kind of leadership, seldom found over such a long period of family domination.

Dexter and Ensign-Bickford stand out as exceptions to the economic laws of mortality, yet they represent the very essence of Yankee prudence and resilience, of those "doers" who—like Sam Colt—were dependent on none but themselves, who flourished by "paddling their own canoe."

CHAPTER IX

The Yankee Genius

The land of stern habits is stretched thro' the middle,
Well known for its sons, scarcely out of their *teens*,
Who will make you to order, all sorts of machines,
From a cotton gin down to a cornstalk fiddle.

DANIEL MARCH, 1840

UNDER the impetus of the Yankee calling, from the very beginning of the U.S. patent system in 1790, Connecticut led all other states in the number of patents granted in proportion to population. Its yearly average, at least until the 1930s, of one patent for every 1,000 residents was three times higher than in any other industrial area of the world. American inventors did not reach their peak until the 1850s, responding in part to more adequate protection of their contributions. The original law provided no control over the merit or originality of a new idea, but in 1836 it was amended to eliminate fraudulent or conflicting claims. This was the year that Samuel Colt received Patent No. 136 for his revolver. The low number assigned Colt reflects the fact that in 1809 the U.S. Patent Office was destroyed by fire, along with all the working models and specifications which had to be submitted with each application. Consequently, no complete account of Connecticut's early inventive activity exists.

In his *Reminiscences* the Hartford capitalist Austin C. Dunham, the principal backer of the Willimantic Linen Company in 1854, credited the patent system for stimulating "the imaginations of great numbers of the citizens of Connecticut and particularly of Hartford. . . . More than two-thirds of the manufacturing industries of Hartford have grown up under the shelter of the United States patent laws,—and without these industries Hartford would indeed be forlorn."

The following tables compare the national trend in patents granted and Connecticut's share:

Period	Average Yearly No. of Patents
1790–1811	77
1820–1830	535
1830–1840	544
1840–1850	646
1850–1860	2,525

Year	U.S. Total	Connecticut
1830	554	52
1840	449	24
1850	973	57
1860	4,510	237

The diversity of inventions from 1790 to 1860 is illustrated by this sampling: Dr. Apollos Kinsley, an improved printing machine (1796); William King and H. Salisbury of Hartford, carriage springs (1804); Ebenezer Jenks of Canaan, fire-brick machine (1808); five patents for different kinds of combs (1809–10); Samuel Green of New London, making paper from seaweed (1809); nine patents for buttons, including three to Ira Ives of Bristol (1815); Eli Terry, thirty-hour wooden clock (1816); four patents to Gilbert Brewster of Norwich for wool machinery improvements (1824); Festus Hayden, wire-eyed buttons (1830); Charles Goodyear, steel spring forks (1831); Edward M. Converse of Southington, wiring machine for tinware (1833); John Howe of Derby, solid head pin-making machine (1841).

A mid-century English observer, Daniel Pigeon, marvelled at the prolific inventiveness of the Connecticut Yankee:

> He is usually a Yankee of Yankees by birth, and of a temperament thoughtful to dreaminess. His natural bent is strongly toward mechanical pursuits, and he finds his way very early in life into the workshop . . . Unlike the English mechanic . . . he cherishes a fixed idea of creating a monopoly in some branch of manufacture by establishing an overwhelming superiority over the methods of production already existing . . . To "get up" a machine, or a series of machines, for this purpose, is his one aim and ambition. If he succeeds, supported by patents and the ready aid which capital gives to promising novelties . . . he may revolutionize an industry, forcing opponents who produce in the old way altogether out of the market, while benefiting the consumer and making his own fortune at the same time. The workshops of Massachusetts, Rhode Island, and especially of Connecticut, are full of such men.

According to Jarvis M. Morse, "the major factor in Connecticut's industrial revolution was the inventive genius of the people." From 1820–1845, over twenty

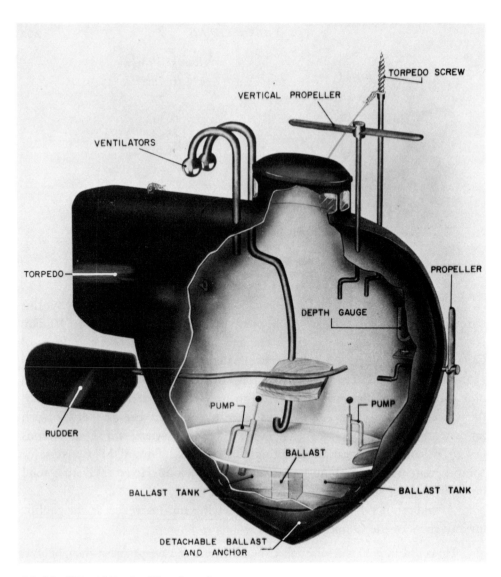

Model of David Bushnell's submarine

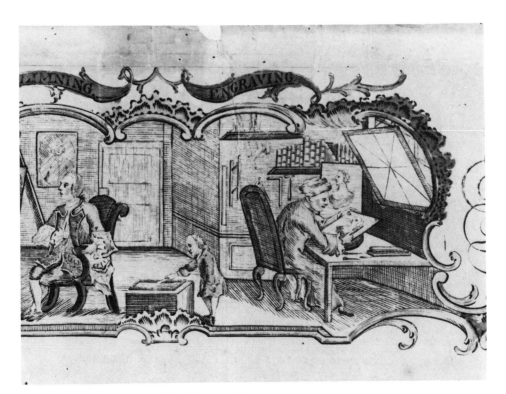

The engraver Abel Buell of Killingworth

John Fitch's steamboat

Fitch plaque at state capitol

patents of major importance were issued to such innovators as Eli Terry, Denison Olmstead, Edward Converse, Samuel Colt, John Howe and Charles Goodyear, most of whom first displayed their talents on the family farm. Several reasons can be cited for this concentration of inventiveness in one small state: the English ancestry of these mechanics, the transformation to manufacturing as the land became exhausted and the maritime trade faded away, the availability of capital from prosperous merchants, the intimate relationship between worker and owner, the burning quest for independence and self-reliance, the innate love of tinkering with tools and machines and the ambition to be the first to make something new.

One of the earliest inventors, who gave birth to a unique defense industry that still flourishes in Connecticut, was David Bushnell of Old Saybrook, acknowledged father of the submarine. His strange, turtlelike craft, built of oaken beams, was launched in 1775, the year he graduated from Yale. Propulsion was supplied by a single brave crewman who sat inside the hull, rowing under water until the air expired about thirty minutes later. A pair of seventy-five-pound "torpedoes" filled with black powder was carried on deck, not for launching but to be screwed to the hull of the enemy ship. He convinced the revolutionary Council of Safety of the practicality of his contraption, the trials went smoothly, but against British warships it thrice failed.

The colorful counterfeiter, Abel Buell of Killingworth, who invented the first gem-cutting or lapidary machine in 1766, as well as a coin press and corn planter, had a long and checkered career, which ended in the almshouse. Basically an engraver, Buell became involved with numerous enterprises in New Haven and Hartford: a lead-type foundry, a packet boat line between New Haven and New London, a stone quarry, and a silverwork shop. After the Revolution he sailed to England in secrecy and illegally brought back a boatload of machinery for a cotton mill that used a power loom.

Eli Terry, the pioneer mass producer of clocks, obtained his first patent in 1797 for a timepiece showing apparent and mean time on the same dial by means of two minute hands operating from the same center. Connecticut's adopted son, Eli Whitney, as related earlier, patented his cotton gin in 1792—an invention that overnight revolutionized Southern agriculture but brought its creator only grief. There was also the unlucky John Fitch of South Windsor, who after a career of peddling, clock repairing and surveying became preoccupied with using steam to move a vessel. His clumsy arrangement of vertical paddles with a reciprocating and lifting motion, something like a canoe, attained a speed of four miles an hour for a distance of forty miles on the Delaware River in 1787, twenty years before Fulton's *Clermont* steamed up the Hudson.

In Hartford lived a peculiar but versatile physician, Dr. Apollos Kinsley, whose fertile mind conceived almost as many different devices as Benjamin Franklin's. Besides a machine for making bricks, he designed three types of cylinder presses and steam engines, two universal pumps—one of which was used on several U.S. warships, three tobacco cutters, a pin machine, a screw-cutting machine, a bullet caster and type caster, a leather currier, a clock with but one cog wheel and a new form of oar. His most intriguing idea was a steam-powered vehicle that actually ran on the streets of Hartford in 1797.

Mrs. Mary Kies of Killingly obtained the first patent issued to a woman, when in 1809 she invented a method of weaving straw with silk thread.

In 1836, Alonzo D. Phillips of Hartford received the first patent for friction matches.

Since the achievements of Connecticut's pioneer gunmakers, clockmakers and waremakers have already been covered, we can turn our attention to the advent of other kinds of manufacturing and to later mechanical geniuses, especially those who followed North and Whitney in the machine tool industry. The latter spurted ahead in the 1850s and, in Connecticut, was concentrated around Colt's great armory in Hartford.

The brewing of coffee attracted several Yankee inventors. Since the early part of the seventeenth century the leading wits, writers and men of affairs had frequented the coffeehouses of London. The pleasant custom of gathering around steaming cups of coffee and a cheery fire was quickly copied in this country. New York's Merchants' Coffeehouse, established in 1737, is claimed by some to have been the birthplace of the American Union. Certainly, this Anglo-Saxon forum served as a catalyst for many enterprises. In Morgan's Coffeehouse, Hartford merchants agreed to share risks on their cargoes to the West Indies and beyond, which led to the founding of the insurance industry and to improved navigation on the Connecticut River. As a substitute for frowned-upon cider, rum or gin, the Yankee switched to drinking endless cups of sugared and cream-rich coffee. Not until 1798, however, did anyone pay attention to a better method for grinding the coffee bean. The second U.S. patent for a coffee mill was issued to Increase Wilson of New London in 1818. A tinmaker by trade, he built a foundry that used the town's first stationary steam engine and prospered to the extent of employing from 125 to 150 men.

The three Parker brothers in Meriden came out in 1832 with an improved version of the cast-iron vertical coffee mill then in general use. Charles Parker, who became Meriden's first mayor and lived to be ninety-two, was the dominant member of the family. For a decade and more the only power in his stone shop was a blind horse that propelled a pole sweep, hour after hour and day after day. Parker was the first Meriden manufacturer to convert to steam. His coffee mills were sold by Yankee peddlers, and, incredibly enough, over the years a hundred different sizes and styles were marketed. Parker ended up managing five large factories that poured out a veritable peddler's cart of products for every taste, including cast-iron vises, gimlet-pointed screws, machine-made spectacles, silver-plated ware, hinges, punching presses, bells, scissors and double-barreled shotguns.

The world's basic inventions have not generally come from scientists or theorists but out of the minds of ordinary pragmatists who at some point in life were struck with a flash of inspiration. The practical inventor par excellence, who by sweat of brow and endless sacrifice, achieved his predetermined goal was Charles Goodyear, the handsome, scholarly, religious, even fanatical discoverer of what Daniel Webster called "elastic metal." Convinced his was a divine mission, Goodyear once explained the agony of invention as follows: ". . . that which is hidden and unknown, and cannot be discovered by scientific research, will most likely be discovered by accident if at all, and by the man who applies himself most perseveringly to the subject . . ." This he did, with far-reaching consequences.

His life almost exactly spanned the period of the Yankee Calling, from 1800 to 1860, and his father's career fitted into the metamorphosis, so common in New England, from merchant to manufacturer. Forced to abandon the West Indies trade when hit by the Embargo of 1807, Goodyear Sr. moved his family from New Haven to Naugatuck to enter the pearl button business. During the War of 1812 he supplied the government with his patented steel pitchforks. At the age of sixteen Charles went off to Philadelphia to learn the hardware trade. When he reached his majority, his father took him into partnership. For a while they prospered, but in 1829 Charles suffered a nervous breakdown—chronic dyspepsia, it was called then—exacerbated by the failure of the hardware store. Like many other small companies, the Goodyears extended too much credit, especially to Southerners, and Charles Goodyear himself ended up in debtors' prison. He accepted his fate with an equanimity that carried him through the rest of his illness-burdened days: "During these years my anticipations of ultimate success never changed, nor were my hopes for a moment depressed."

Though down and out, in delicate health, without any technical education, he determined to make inventing his profession and, further, to find a way of making gum elastic, or India rubber, a useful product for mankind. Rubber became his passion, mistress and Holy Grail. In 1820 the importation of Brazilian rubber shoes created tremendous excitement when they landed in Boston. Several enterprises for making rubber goods were launched. Goodyear read with avid interest about the tribulations of Edwin Chaffee in Roxbury, who tried to make leather shoes waterproof by applying a gum covering. Unfortunately, the unstable material softened and melted in hot weather and cracked in winter, baffling the chemists and enraging the wearers. A chance visit to a New York store in 1834, where he saw a rubber life preserver in the window and found it worthless, impelled Goodyear to embark on his long series of experiments. Strangely enough, the revelation that came five years later proved that only a high degree of heat, in combination with sulphur, would make the natural gum pliable and indestructible. Goodyear learned about the effect of sulphur on rubber from Nathaniel Hayward, a foreman in a defunct rubber factory, who later started the Hayward Rubber Company in Colchester (1847) and had a second factory for preparing crude rubber at Bozrahville.

Goodyear carried on his research, which required only a few low-cost materials to pursue, in several places, beginning in debtors' prison in Philadelphia and continuing in New Haven, New York, Springfield and

Charles Goodyear

Roxbury. The "wonderful elasticity" of rubber never ceased to fascinate him: "It can be extended," he said, "eight times its normal length . . . when it will again assume its original form. There is probably no other inert substance, the properties of which excite in the human mind an equal amount of curiosity, surprise and admiration . . ." He patented his ideas for thin rubber sheets and a curing process. When someone in New York asked how he might recognize the inventor, the answer was: "If you meet a man who has on an India rubber cap, stock, coat, vest and shoes, with an India rubber money purse without a cent of money in it, that is he." Just as he seemed on the verge of success, with plans to start manufacturing rubber goods on a large scale, the panic of 1837 knocked him flat, but temporarily he was saved by another patent for rubber shoes.

One winter evening in 1839, while working in the kitchen of his Woburn, Massachusetts home, a piece of gum mixed with sulphur in his hand accidentally touched the hot stove. To his amazement "it charred like leather without dissolving. No portion of it was sticky." He nailed it outside the door in the intense cold, and the next morning was delighted to find it absolutely flexible. In this one dramatic moment he had learned to cure rubber all the way through so as to resist cold, heat and acid and, in effect, to create an entirely new material. It took him two more years to convince anyone else of the value of his discovery. Ridiculed, destitute, highly nervous, emaciated and sallow, with a family of young children to support, he had no place to turn for help. His friends had grown tired of his reckless obstinacy; his vulcanized rubber looked exactly as it always had. So back to debtors' prison he went, until finally his brother-in-law, a Naugatuck wool manufacturer, and a few others advanced him the necessary capital to get his patent in 1844—ten years after his torturous adventure began. His brothers joined forces with him, Nelson contributing the process for making hard rubber and Henry helping to establish a rubber shoe plant in Naugatuck that used steam to dissolve the gum. William DeForest, the brother-in-law, was also involved in forming not only the Goodyear Metallic Rubber Shoe Company but also the Naugatuck India Rubber Company, which, together, were turning out rubber goods at the rate of $120,000 annually by 1845.

Back in New Haven, Charles eked out a living from the sale of licenses. At first he received a royalty of three cents per pair of shoes, but it was soon reduced only to one half-cent. A poor businessman, he never collected more than a fraction of what was due him. He published a pamphlet listing over fifty types of articles in which he believed "metallic gum elastic" would surpass any other material. Soon, to protect themselves against a rash of patent infringements, his license-holders joined forces and hired Daniel Webster for the whopping fee of $15,000—more than Goodyear earned from his invention altogether. To clear his claim took seven years of court battles involving more than sixty suits.

A number of rubber companies carried on the Goodyear name in Connecticut, including L. Candee, Seamless Rubber and—largest of all—U.S. Rubber, a combine organized in 1892. At the world's first exhibition of progress held in 1851 in London's Crystal Palace, Goodyear won a medal and was acclaimed as a benefactor to mankind. The only other Connecticut Yankee in attendance, who also created quite a stir, was the flamboyant Sam Colt. Never able to escape the spectre of debt, even when journeying abroad, the exhausted but still dedicated Goodyear could say on his deathbed in New York: "What am I? To God be all the glory."

If the clock was the first mass-produced thing to become a household necessity, then the sewing machine came second. The original model appeared in England as early as 1790, a chain-stitch machine patented by one Thomas Saint. But for sixty years his idea gathered dust, and it remained for American inventors to resurrect it. A robust, curly-headed country boy from Massachusetts, without any knowledge of prior efforts abroad, conceived a machine, as a result of his cotton mill experience, embodying a curved eye-pointed needle and a lock or weaving stitch. It worked on the same principle as a shuttle in a loom, but lacked a suitable feeding device. In 1843, this young man, Elias Howe, came to New Hartford to work in Greenwood's cotton mill. In the basement of the hotel where he lived, he perfected his sewing machine. The story is told that he solved the feeding problem through a dream in which he was captured by savages and dragged before their king:

> The king issued a royal ultimatum. If Howe did not produce a machine that would sew within 24 hours, he would die by the spear. Howe failed to meet the deadline and saw the savages approaching. The spears slowly rose and then started to descend. Howe forgot his fear as he noticed that the spears all had eye-shaped holes in their tips. He awakened and realized that the eye of his sewing machine needle should be near the point, not at the top or in the middle. Rushing to his work bench, he filed a needle to the proper size, drilled a hole near its tip and inserted it. It worked.[1]

His 1846 patent brought him no reward until near the end of his life. Enraged by the success of other manufacturers, like Isaac Singer, he filed a number of suits for patent infringement. In 1854, Singer was forced to pay $28,000 in settlement of Howe's claims. Four years before his death in 1867, Howe was able to build a large factory in Bridgeport with special equipment for each operation to insure accuracy and uniformity in every part.

About the same time as Howe struggled to get on his feet, Allen B. Wilson

---

[1] *Psychology Today*, June, 1970.

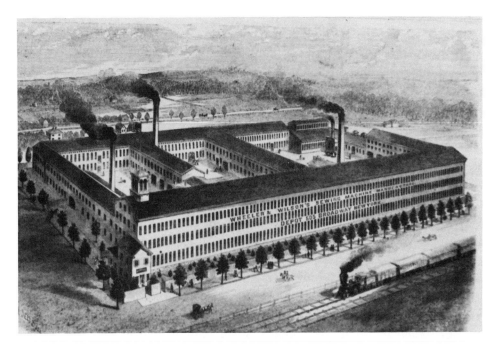

Wheeler & Wilson's sewing machine manufactory, Bridgeport

Jewell Belting Company and Park River, Hartford

devised an even more practical sewing machine that could sew continuous seams of any length, straight or curved, and turn corners at any angle. He received two patents, one in 1850 and the other a year later, which happened to fall on the same date as Isaac Singer's patent. Wilson formed a partnership with Nathaniel Wheeler, a manufacturer of small metalwares in Watertown. Together, they took Wilson's machine to Oliver Winchester, later the founder of Winchester Repeating Arms, but then a shirtmaker in New Haven. At first Winchester refused to try "the contrivance." But when Mrs. Wilson stitched a shirt perfectly in his presence, he jumped at the chance to acquire the patent rights for New Haven County. Soon Wilson machines were introduced to shops in other cities. In 1856, Wheeler and Wilson moved to Bridgeport. Their plant, previously occupied by the defunct Jerome Clock Company, quickly grew to be the largest of its kind in the world, lighted at night by 600 gas burners. In 1860 alone, it turned out 19,265 sewing machines. Nearly a half century later the company merged with Singer. Wilson's eminently practical machine, with its ingenious rotating hook and stationary bobbin, made him well-to-do and the sewing machine as common a family possession as the clock on the mantelpiece.

No man could have had a more dichotomous name for a gunmaker than Christian Sharps, who came from Cincinnati to Hartford in 1851, just as his rival Sam Colt was buying land in the South Meadows for his gigantic armory. Indoctrinated into the Whitney system of interchangeable parts by Captain John H. Hall at Harper's Ferry and inspired by Hall's breech-loading rifle, he designed and patented a much simpler model that was destined to become one of the most famous single-shot firearms. It was possible to load and fire it five times a minute. During the Civil War some 100,000 Sharps rifles and carbines replaced the old muzzle loaders; many more accompanied every wagon train that pushed westward. It was the favorite weapon of buffalo hunters. In 1850, Sharps brought his invention to Robbins & Lawrence in Windsor, Vermont, a small metal-working company whose management exerted a tremendous impact on machine design and development. They assembled 5,000 cavalry carbines with standardized parts. When a contract for 10,000 more was negotiated the next year, the Vermont firm, having run out of space, chose to build a separate plant in the better-located manufacturing center of Hartford. The large tract of pasture land near the Little River that was selected developed into a spawning ground for such famous companies as Weed Sewing Machine, Hartford Machine Screw, Pratt & Whitney, United Aircraft, Arrow-Hart, and Underwood Typewriter. Christian Sharps went along as chief engineer but stayed only two years, since his partners found he lacked production ability. The two-story brick armory, with a forge shop adjoining, was of imposing size for that day, measuring 160 by 60 feet.

Soon 250 mechanics kept the Windsor-built millers and multiple-spindle drill presses running.

Six years later Robbins & Lawrence failed, primarily as the result of the cancellation of a large English rifle contract in connection with the Crimean War. The Sharps Rifle Manufacturing Company, suddenly cast adrift, now had to make its own machines. Richard S. Lawrence, who remained as master armorer, told how his company passed up the opportunity to become one of the great machine tool builders in the country: "We had run a regular machine shop at Windsor, and we continued this work at Hartford, making most of the machines used in our factory, many for the English and Spanish governments, and for other gun and sewing-machine builders. I tried to have the Sharps Co. enter into this business more extensively, as there were bright prospects for the future, but they declined, and this is what brought Pratt & Whitney into the business."

Beginning in 1855, Sharps carbines were bought by New England abolitionists and shipped to antislavery forces in the Kansas Territory, where in the hands of fighters like John Brown, a Connecticut son, they were a decisive factor in making it a free state. Brown also used them in his Harpers Ferry raid. Another militant abolitionist, the Brooklyn preacher Henry Ward Beecher, whose sisters lived in Hartford, claimed there was more "moral power" in a Sharps rifle so far as the slaveholders were concerned than in a hundred Bibles. The cases in which the guns were packed were sometimes marked "Bibles," and soon were called "Beecher's Bibles."

Another pioneer lured to Hartford during this period contributed the means for operating a multitude of machines simultaneously from one source of power. In 1845, Pliny Jewell, a native of Winchester, New Hampshire, whose family had engaged in the art of tanning for several generations, set up a tanyard near the Little River. Three years later he began to specialize in making leather belts up to forty inches wide as an economical substitute for the costly and cumbersome system of gearing then common in factories. Most of the leather came from a tannery which Jewell built in Detroit. One of his four sons, Marshall, served three terms as governor of Connecticut and as ambassador to Russia, where he discovered the process of making scented leather.

Certainly the most durable Connecticut inventor, perhaps its most prolific and brilliant one, was a quiet little Yankee named Christopher Miner Spencer. Born in Manchester in 1833, he died in Hartford ninety years later, a span that included separate careers in three different industries: silk, firearms and machine tools. The son of a wool dealer, Spencer went to work at the age of fourteen in the Cheney silk mill; during his spare time he constructed a working model of a steam engine based on information gleaned from an old textbook, *Comstock's*

Christopher Spencer

Philosophy. In 1853, he invented an automatic silk-spooling machine that released hundreds of girls from the tedious task of manual spool winding. After taking time out to finish his high-school education, acquire his journeyman's certificate and learn more about toolmaking, the Colt charisma ensnared him for a period of two years. In 1856, Spencer returned to the Cheneys as superintendent of their machine shop. One of his inventions, a steam carriage (1862)—like Dr. Kinsley's—was way ahead of its time. To a plain four-wheeled buggy he added a boiler and two-cylinder engine with a chain drive to the rear axle. It could run forward only, the direction he always wanted to go, and for a while he drove it to and from work. On a local race track his vehicle could keep pace with the fastest trotters.

For three years Spencer worked on a seven-shot rifle on his own time after an eleven-hour stint in the mill. Colonel Frank Cheney encouraged him to obtain a patent, which was issued in 1860. Commented the centennial edition of the *Manchester Herald* the year of Spencer's death: "If personal credit for winning (the Civil) War could be granted, a large share would fall to . . . Spencer, whose invention of an automatic repeating rifle was a contributing factor towards the success of the Union Troops . . ." At the start of hostilities the Cheneys set up a separate company in Boston which produced more than 200,000 Spencer rifles, initially for Connecticut, Massachusetts and Michigan volunteers. The inventor was paid a royalty of one dollar per gun, which could fire 21 shots in a little more than a minute. At the end of the War Winchester Arms acquired the equipment of Spencer Repeating Rifle Company but, according to Spencer, "squelched the Spencer rifle which had been a formidable competitor."

Undoubtedly, Christopher Spencer's greatest invention was the automatic turret lathe or screw machine. While at Billings & Spencer he patented a machine for automatically turning the spindles and heads of metal spools used in sewing machine shuttles, thereby reducing the cost eighty percent. In turn, this device suggested to his restless mind the screw machine. So confident was he of its practicability that he secretly made a wooden model and applied for a patent before he had even built a prototype. For his first working model he adapted a Pratt & Whitney hand-operated turret lathe. On capital supplied by George Fairfield the Hartford Machine Screw Company began in 1876. Still later he invented a five-spindle screw machine which created the Universal Machine Screw Company, also in Hartford, with which the never-say-die inventor remained actively connected until 1912.

The name Pratt & Whitney conjures up an image of jet engines more than one of intricate machine tools and precision gages. Yet the latter were essential precursors to the emergence of the automobile and aircraft industries. The Pratt & Whitney Company, founded by Francis A. Pratt and Amos Whitney in 1860, has made enormous contributions to the metal technology of the world. Amos Whitney, a distant relative of Eli, left his home in Lawrence, Massachusetts, as a mere lad of fifteen to seek his fortune in Colt's "college of mechanics." From Lowell, in 1852, came the older Pratt to work under the great Root. Within the Colt Armory a strong bond of friendship developed between the two young machinists. Simultaneously, the Vermont inventor Richard Lawrence brought to the Sharps Armory in Hartford the new milling machine designed by Frederick W. Howe at the Robbins & Lawrence plant, which represented a basic improvement over Eli Whitney's.

Two years later, Pratt took over as plant superintendent at the George F. Lincoln & Company, formerly the Phoenix Iron Works and later Taylor & Fenn. Whitney soon followed him as a contractor. In 1855, Pratt copied the Howe design but substituted a screw feed for the rack and pinion, thus creating the Lincoln plain miller, which became the most popular machine tool of its kind. Over 200,000 were built. In fact, the Lincoln or Pratt miller formed the genesis of Pratt & Whitney's machine tool business after the Civil War.

In a small room which they rented on Hartford's Potter Street, Pratt & Whitney entered into partnership as "moonlighters." The only furniture was a stove. Throughout the Civil War they kept their jobs at Lincoln and in their spare time did machine work for others. One of their first accomplishments was building Christopher Spencer's automatic silk winders for the Cheney mills, later adopted by the Willimantic Linen Company.

In less than forty years, machine tool building, under the stimulus of men like Pratt and Whitney, grew into a lusty new industry centered mostly in Hartford; Providence; Windsor, Vermont; Manchester and Nashua, New Hampshire. Lowell and Philadelphia dominated the manufacture of textile machinery. Nevertheless, machine tools did not exist as a separate industry until well after the Civil War, evolving out of a complex series of close relationships between numerous inventors, companies and products.

What hands are to a man, chucks are to a machine—holding devices for a piece of work and the tool that shapes or drills it. The antecedent of the modern lathe chuck was the potter's wheel. Surprisingly, early English mechanics overlooked the necessity of adequate tool or work holders for their machines, and the first to do something about them was a skilled loom repairman in a West Stafford machine shop. In the bogs of this area, iron ore had been uncovered about 1800, giving rise to metal-working shops that kept busy casting cannon during the War of 1812. Frustrated by the failure of oak blocks screwed to the face plate of his lathe to hold the work he wanted to machine, Simon Fairman devised a crude, geared scroll chuck of iron. On July 18, 1830, President Andrew Jackson signed Fairman's patent letter for an "Expanding and Contracting Universal Chuck for Lathes." Six years later Fairman left his employers to build his own shop in West Stafford. Two of his employees—John H. Washburn and Austin F. Cushman—also became his sons-in-law. To expand his market, Fairman moved to Watertown, New York, and began peddling chucks from one shop to another. Another employee, Eli Horton, broke away in 1851 and started making a universal chuck in Windsor Locks; later, David Whiton, who had become an active partner, set up a similar business in New London. At the outbreak of the Civil War Cushman went to work in the booming Colt Armory,

Amos Whitney as a 14 year-old apprentice

but soon decided to launch his own chuck enterprise. In the bedroom of his home, using a simple lathe operated by a foot treadle, he laboriously turned out a new version of the self-centering chuck based on his father-in-law's patent. This was the beginning of the Cushman Chuck Company in Hartford, which in time acquired most of the other early chuck companies and became the undisputed leader in its field.

One day in 1851 the local paper in New Haven reported that Eli Whitney Blake had been appointed to a committee to construct two miles of macadam highway leading to the village of Westville. The state's first match company had just been started there by A. Beecher & Sons, while Blake ran a lock factory nearby. At that time there were not a dozen miles of macadam road in all of New England. Stone for road building had to be prepared by hand, a painfully slow application of manual labor still prevalent in underdeveloped countries. Blake observed: "It took two days' labor to produce one cubic yard of road metal." So he abandoned his other interests to devise a mechanical stone-breaker for crushing the sturdy trap rock of Connecticut hills. After seven years of the most thorough experimentation he was ready for a patent in 1858.

A Yale graduate and nephew of Eli Whitney, Blake had worked in his uncle's armory and shared a similar inventive turn of mind. After managing the armory for a decade, until 1835, he started the lock and hardware business with his brother Philo in Westville, introducing the mortised lock in place of the old box type. From the beginning his stone crusher operated perfectly and had an immediate impact on all ore crushing and civil engineering. It could reduce the largest stones to the required size, applying a pressure of 27,000 pounds to the square inch. It has been called as basic an invention as Morse's telegraph, McCormick's reaper or for that matter Eli Whitney's cotton gin, since it made possible the development of reinforced concrete and the building of highways, railroads and dams. Yet Blake is seldom acknowledged in the list of America's great inventors.

Clocks and springs, those almost magical devices that store or exert energy, are so closely related as to be kissing cousins. When the Bristol clock industry was enjoying its heyday, Col. E. L. Dunbar, nephew of the clockmaker Silas Hoadley, having either acquired or invented a method of tempering steel, began making springs. A decade later the arrival of crinoline skirts in the fashion world inspired him to make hoops out of his spring steel to keep them nice and stiff. Another Bristol mechanic, the jovial Wallace Barnes, through a clever series of trades quite unexpectedly found himself owning the business for which he

Francis Pratt

Amos Whitney

Pratt & Whitney Company, Capitol Avenue, Hartford, 1870

First Pratt & Whitney machine tools, 1860. Lincoln-type miller

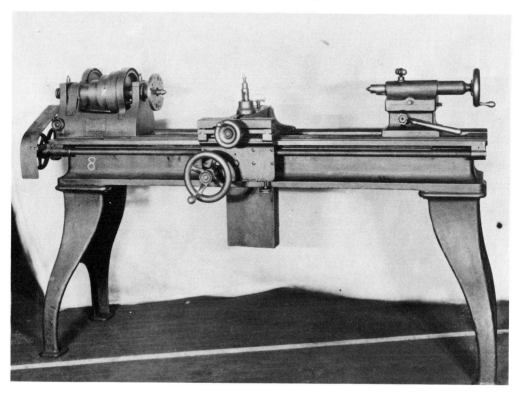

Lathe with weighted carriage

Pratt & Whitney complex, 1930

First chuck factory, West Stafford, 1845–54

worked, as one good result of the depression of 1857. This was the A. S. Platt & Company, which also made hoops and clocks as well.

Descended from a family of traders and storekeepers, Wallace Barnes advertised himself as willing to trade for anything under the sun. He raised fruit trees, speculated in real estate, bred Jersey cattle, but found his great opportunity soon after he began assembling clock alarms under contract to A. S. Platt. When Platt went bankrupt, in lieu of back pay, Barnes was given a quantity of hoop wire, which he loaded onto a wagon and drove to Albany. According to a descendant, "In rapid succession and with the genius of a true trader, he bartered the wire for a haberdashery—swapped the store sight unseen for a farm in Missouri where he had never been—exchanged this for a blacksmith shop in his home town—sold that for $1,600 and bought the plant from which he had received the wire in the first place!"

Barnes and Dunbar then merged their businesses and built a factory appropriately called "Crinoline Hall," which for a while ran three eight-hour shifts—an unusual practice in those days. The coiled steel strips were shipped in barrels to New York, where a bevy of young ladies completed the skirts. The hoop skirt fad evaporated during the Civil War, and the partners turned again to making springs for the more dependable clock market and later for muskets and powder horns. From that time forward the spring industry found a multitude of applications throughout the spectrum of industrial products.

It has often been stated that Yankee ingenuity, adaptability and initiative have been Connecticut's greatest resource. Certainly, Connecticut can claim credit for the origin of more basic inventions and more important industries than any other state. Even before the Civil War it produced an astounding diversity of useful products that eased the average man's life. By that time Americans everywhere, thanks to the peripatetic peddler and the iron horse, could wake up in the morning to the striking of brass clocks made in Bristol, Thomaston, New Haven or Waterbury. They stepped on carpets woven in Thompsonville; softened their beards with Williams shaving soap from Glastonbury; put on clothes made from Somersville wool, Rockville cassimere, Norwich cotton or Manchester silk, fastened with Waterbury buttons, hooks and eyes, and sewed with Willimantic thread—including a New Haven shirt, a Bridgeport corset, a Meriden or Bristol hoop skirt, and cotton hose and suspenders from Middletown; dined on silverware from Meriden or tinware from Berlin, using a Parker coffee mill to grind the beans for their brew; sat in Hitchcock chairs; celebrated with a swig of Enfield cider brandy; smoked a cigar with a Windsor wrapper; cleaned their teeth with Comstock ivory toothpicks; put on a Danbury hat and a

Cushman Chuck's work force in the 1860's

Naugatuck pair of rubbers; turned the key in a Blake or Yale lock; rode on a Hartford saddle or in a New Haven carriage jingling with East Hampton sleigh bells on streets macadamized (at least in Connecticut) with native trap rock; sailed in Mystic-built ships; hunted with a Colt revolver or Sharps rifle, using Hazardville powder; and built houses with facings of Portland brownstone. With the help of Howe's common pins housewives made new clothes on the Wilson sewing machine; their children looked up words in Noah Webster's *Dictionary* printed in Hartford under a Waterbury kerosene or whale oil brass lamp lit with a Beecher friction match; on the wall hung pictures made from Waterbury daguerrotype plates. Daughters prettied their hair with Essex ivory combs and played on applewood flutes made in Fluteville. Gardens were planted with seeds packaged by Enfield Shakers and fields tilled with a Wethersfield plow or Higganum hoe; canning was done with Lyman's patented, air-tight fruit jars; sheep were sheared or butchered with Hotchkissville blades; mice were disposed of by a Bostwick mouse trap from Sharon; while fathers fished with Jenkins's spring-steel fish-hooks, blasted with Ensign-Bickford's safety fuse and cleared the forest with a Collinsville axe or machete.

Eli Whitney Blake, 1795–1886

Blake's stone crusher

FACTORY OF THE SWIFT & COURTNEY & BEECHER CO., AT WESTVILLE, CONN.

CHAPTER X

Manufacturing First

HOWEVER prolific its inventions, however indefatigable its peddlers, however voluminous its production of guns, clocks and other Yankee wares, Connecticut was still not predominantly a manufacturing state at mid-century. Fewer than twenty percent of its people lived in towns of more than 5,000. Approximately 30,000 worked in mills and shops. Most manufacturers, although they had adopted the factory system to one degree or another, employed only fifteen to twenty-five workers. The two largest factories were the Collins axe and Thompsonville carpet companies. Enfield's distilleries had disappeared. Metalworking, nevertheless, was growing rapidly, as these figures show:

Product	Number of Factories	Centered In
Tinware	101	Hartford to New Haven
Brass	48	Waterbury
Buttons	42	Waterbury
Clocks	30	Bristol
Plated Silverware	13	Meriden

By Town		Employees (5 or more)
Hartford	145	2,400
New Haven	37	2,137
Waterbury	34	787
New London	13	179
10-town Total	431	10,300

Employers were more aware than ever that people problems could be as demanding as those of machines, markets, credit and profits. In his inaugural address in 1842, Governor Chauncey Cleveland, a Democrat, showed concern over the tendency for factory owners to interfere with the voting rights of workers. Too frequently, it seemed, the threat of discharge hung over the head of the misguided employee who failed to vote as his employer thought. For example, the "Federalist" Henry Thompson, the Thompsonville carpetmaker, was accused of controlling the franchise of his workmen. His public denial included a statement signed by thirteen of his weavers to the effect that as supporters of Jackson's political principles they had never been intimidated from doing as they thought right. The *Hartford Times*, a Democratic paper, refused to print it.

Cleveland, moreover, deplored the long hours children were required to work in factories. The legislature obliged by prohibiting the employment of those under fourteen years of age more than ten hours a day but limited the restriction to cotton and woolen mills. As long as a child was over fourteen, apparently, he was considered strong enough to put in twelve hours of labor.

In Hartford County at this time the leading industries were carpeting, paper, boots and shoes, clocks, hosiery and book publishing. As early as 1820, twenty publishing houses flourished in Hartford. In 1836, Newton Case had founded a printing firm that still exists under the name of Connecticut Printers. His success came from being the sole publisher for fifteen years of Webster's *Dictionary*, from *The Cottage Bible* and from school books. There were only three machine shops in Hartford, employing forty-five men. Colt's great armory was yet to come, but when it did around 1853, it marked the beginning of Hartford as a manufacturing center, especially in metal-working.

Better transportation tied the industrial communities to Boston and New York, both centers for distribution agents and bank credit. By 1850 the railroad ran all the way from Hartford to New York and even reached Waterbury, thus speeding up Connecticut's transition to a manufacturing economy. The main indicator of what was happening was the loss by agriculture between 1840–50 of 25,000 workers and its decline in importance from sixty-one percent to only one-third of the state's total income.

Right up until 1860, Connecticut's leading industry—in terms of both employment and value of output—remained cottons and woolens, the manufacture of which was concentrated, respectively, in Norwich and Rockville. Weaving was no longer done in the homes. Mills diversified into cotton drills, shirtings, prints, ginghams, woolen broadcloth, cassimeres, flannel, blanketing and tweeds. Since its pioneering, however, the textile industry had contributed almost nothing to the state's overall industrial progress. The real milestones were being achieved by the dramatic efforts of the mass producers, those who had successfully applied the techniques of standardization to such products as guns, clocks and sewing machines. Except for the Thames Manufacturing Company of Norwich and the Windham Manufacturing Company in Willimantic, each of

Case, Lockwood & Company, Hartford

Rockville Woolen Mill

which produced better than a million yards of cotton cloth a year, the leadership in textiles had long since passed out-of-state to Lowell, Waltham and Pawtucket.

Massachusetts alone made four times as much cotton cloth and prospered exceedingly. In most Connecticut mills, which averaged twenty or fewer hands, earnings were paltry by comparison. In fact, Connecticut capitalists showed surprising indifference to the cotton industry and allowed Rhode Island interests to move into the southeastern part of the state and to operate subsidiaries. The largest cotton mill in 1843, the Quinebaug Company, was owned by the Lippett brothers of Woonsocket. Governor William Sprague, acknowledged as the preeminent Rhode Island cotton manufacturer, initiated a large-scale expansion in Connecticut in the late fifties. Outside of Norwich, along the Shetucket River, he bought a tract of 8,000 acres and constructed an enormous mill of gray stone nearly a thousand feet long, four stories high, with six waterwheels that developed 950 horsepower. The mill had a capacity of one million yards of cloth per month. Around it he laid out a village for the projected employment of 1,300, that included 100 cottages, a company store, school, church and town hall.

Massachusetts also dominated the woolen industry; it grabbed the sizable ivory comb market away from Meriden; and through mechanization became the footware center of the country. A decade or so earlier Connecticut's second largest industry, one carried on in nearly every town, had been boots and shoes, but it never overcame its strictly household or manual status and quickly faded away. Notwithstanding the competition of its sister states, at the start of the Civil War, Connecticut ranked fifth in cotton output, with 129 mills, and third in woolen goods, with 84 mills.

* * *

The year 1860 was momentous for Connecticut manufacturing. The slavery issue neared a nation-splitting climax. Over the fate of the Union "Black Republicans" diametrically opposed Democrats, who on this subject had taken a surprisingly conservative stance. Although William Buckingham of Norwich was narrowly reelected governor on the Republican ticket and Lincoln spoke reassuringly in three towns during his successful campaign, the state's manufacturers themselves became a house divided. Buckingham himself doubled in brass as the well-to-do president of the Hayward Rubber Company in Colchester and during the war headed a syndicate of Norwich businesses anxious to capitalize on the "cotton famine." Conflict of interest was no problem to them. The depression of 1857, which had retarded the output of textiles and hats, had been forgotten but the seven percent interest rate for loans hamstrung and annoyed Yankee businessmen. Most of their attention, in January anyway, was directed at the upcoming annual convention of manufacturers in Meriden, which

Workers' homes next to American Thread Mill, Willimantic

MANUFACTORY OF THE WILLIMANTIC LINEN CO., WILLIMANTIC, CONN.

Willimantic Linen Company, 1857 (later The American Thread Company).

promised to be a hot forum for debate and decision on where they stood in regard to slavery and Southern customers.

Many manufacturers, like Samuel Colt, were Democrats who believed the North's interference in the South would mean the end of the free trade upon which the state's prosperity depended. They dreaded the prospect of breaking up the nation as well as their party and losing their lucrative markets in the South. New Haven carriages as well as Colt, Sharps and Whitney guns were earning fat profits. Colt had received a $50,000 order to reequip Virginia's militia. Other manufacturers were moderate Republicans who favored a strong protection policy and, though not in favor of abolition, did object to the extension of slavery. The Meriden conclave drew representatives from some eighty leading firms, the blue bloods of Connecticut industry, among whom were:

Samuel W. Collins, Collinsville Axe Company
Raymond French, Humphreysville Manufacturing (tools)
George Goodyear, Beacon Falls Rubber Company
W. H. Perry, Wheeler & Wilson Sewing Machine Co., Bridgeport
Dennis C. Wilcox, Meriden Britannia Company
William Wilcox of Middletown (marine hardware)
Aaron & W. M. Pratt of Meriden (ivory combs)
Henry & James Terry of Plymouth (clockmakers)
John C. Palmer, president of Sharps Rifle Co., Hartford
Charles Parker of Meriden (hardware)
H. A. & Samuel Yale of Meriden (tinware)
Julius Hotchkiss, Russell Manufacturing Co., Middletown
George M. Landers of New Britain (cabinet hardware)
O. B. North of New Britain (saddlery hardware)
H. E. Russell of New Britain (hardware)
R. E. Hitchcock of Waterbury (buttons)
R. A. Coe of Waterbury (brass)
E. H. Plant of New Haven (carriage bolts), and
Six carriage builders from the same town

In addition to the "regulars," who had called the meeting, an articulate group of "outsiders" attended, most of them merchants and stockholders from Hartford who were all set to carry the day for the Republican viewpoint. One of their leaders was J. Pratt Allyn, a fervent Abolitionist, who spuriously claimed he represented the Hartford Carpet Company, the "biggest corporation in the state—over $1,500,000," he said. Democratic sympathizers, believing they were going to pass resolutions backing the Constitution and the Union, instead found themselves outvoted, or at least outshouted, by Republicans. One of the latter, Calvin Day of Hartford, a director of three companies, called them "greasy mechanics." The resolution which the Republicans offered seemed innocuous enough:

> That as manufacturers, we know no North, no South, no East, no West in the sale of our merchandise, only the Union . . .

Gagged by the "bogus" manufacturers, the regulars withdrew and met separately in a room below the hall, declaring their opponents had insulted the South and were apologists for John Brown. They soon adjourned "with three rousing cheers for Washington, the Union, and equal and exact justice to North and South." Concluded the Republican *Courant*:

> . . . The bombastic bluffers for the South . . . expected to make capital, by frightening the manufacturers of this State into the idea that nobody would buy their wares, if they did not emasculate themselves of their political manhood . . . The Meriden Convention has worked gloriously for the Republicans . . . "Jim Babcock and the Asylum Street merchants" did the State good service, in going to Meriden and just upsetting the nice little family arrangement of persons in the Shamocratic interest for frightening timid voters from their Republican faith.

Colt, for one, was not silenced. During the state campaign that followed, his political convictions caused a great stir in the local newspapers, the *Courant* leading the attack and Burr's *Times* waging a vigorous defense. Without doubt in previous Hartford elections Colt had used persuasive tactics, including ballot box watching, to have his men support Democratic candidates for aldermen, especially Mayor Henry C. Deming. This time he was accused of discharging outright "66 men, of whom 56 are Republicans . . . Many of these were contractors and among his oldest and ablest workmen." The *Evening Press* pointed out that most of those fired lived in the Fourth Ward where the Armory stood. Asserting that their dismissals amounted to proscription for political opinion, Colt's Republican workers resolved at a meeting that "the oppression of free labor by capital, and the attempt to coerce and control the votes of free men, is an outrage upon the rights of the laboring classes." Colt quickly issued a flat denial:

> In no case have I ever hired an operative or discharged one for his political or religious opinions. I hire them for ten hours labor . . . and for that I pay them punctually every month . . .

Yet a few months earlier he had suggested to a politician that he pen a resolution urging all employers to discharge every "Black Republican" until the slavery question was set at rest and the rights of the South secured. And in the spring of 1861 from Cuba, where he was taking one of his infrequent vacations, Colt exhorted his superintendents, Root and Lord, to run the Armory night and day with a double set of hands in order to "make hay while the sun shines."

When the first shot was fired at Fort Sumter, the nation's most famous and affluent gunmaker was nearing the end of his meteoric saga. Christopher Spencer had just patented his rapid-fire rifle. Young Pratt and Whitney had launched their machine tool and gage business. Wallace Barnes waxed rich on his hoop skirts for ladies. Samuel Collins did likewise from his outpouring of machetes for the foreign market. Eli Whitney Blake had just revolutionized the building of dams and roads with his stone crusher. Frank W. Cheney prepared to return from China and Japan, where he had been buying raw silk for the family mills in South Manchester. Salisbury furnaces were busy forging cannon. Goodspeed launched ships at East Haddam, as runaway slaves were being smuggled upriver on steamboats. Eli Terry and Charles Goodyear—two of Connecticut's greatest inventors—had recently died. Chauncey Jerome, the great clock manufacturer in New Haven, was penniless.

In a number of industries Connecticut had established leadership. New Haven's carriage builders, employing over 1,800 men, accounted for over two-thirds of all the axles being forged. Half of the nation's firearms originated from Hartford and New Haven; more than half of all brass buttons and general hardware from Waterbury and New Britain; most of the clocks, common pins, India rubber goods, silverware and rolled brass from other towns. Tinware, on the other hand, had shrunk in importance. Manchester was the leading silk producer. Besides 41 carriage shops, New Haven had 175 factories employing over 6,000. Nearly a hundred more hummed in Bridgeport, with carriages and sewing machines the major products. Wheeler & Wilson was the largest sewing machine factory anywhere, capable of producing 20,000 machines or better a year. Around Hartford, besides Colt pistols and Sharps rifles, the chief manufactures were carpets, cigars, hardware, paper, sewing silk, textiles, axes, gunpowder, hosiery, silver plate and Britannia ware. In Danbury, a thousand men and women made $1,500,000 worth of fur and wool hats, the third-largest volume from any state. And 534 Yankee peddlers still hawked Connecticut wares far and wide.

Thus, in 1860, the state which the South and its cotton economy had propelled into becoming a manufacturing giant was now ready to use its brawn to crush it.

* * *

Forty percent of the gunpowder consumed during the Civil War came from Powder Hollow on the Scantic River. In fact, the powder used by Confederate artillery in Charleston against Fort Sumter was Connecticut made. Appropriately enough, the village was called Hazardville—not because of the nature of the product, but after its founder, Colonel Augustus G. Hazard. His gigantic operation, covering 400 acres and including over a hundred separate mills, twenty-five water wheels, three dams, miles of canal and a narrow gauge railroad, made him one of the Big Three in powdermaking. This complex could barrel as much as twelve tons of powder daily, equal to more than a million dollars annually.

A Yankee peddler from Suffield, Allen Loomis, tired of bartering cigars, whips and powder for furs from trappers in Northern New England, started the business in 1835. He formed a partnership with Allen A. Denslow of New Haven and two brothers and sent to Kent, England for a couple of experienced hands and their families. Nine more powdermakers migrated, stayed only a short time and moved on to become Ohio farmers. But the real dynamo was Colonel Hazard, a New York merchant and ship owner who had grown up in Columbia, Connecticut. Hazard had done well selling paints and oils in Georgia and investing in a packet line from New York to Savannah. As agent for Loomis & Denslow, he also invested in their new business. After losing his ships in the panic of '37, Hazard bought out the Loomises and organized his own joint stock company in 1843.

Soon he built a handsome twenty-seven-room mansion in Enfield near the Congregational church where Jonathan Edwards had preached his famous sermon during the Great Awakening a century earlier. Modeled after the gracious plantation estates in the South, of French colonial style, the house boasted a solid cherry floating staircase that cost its owner $40,000. Somewhat to the consternation of his fellow townsmen he held numerous gay parties and, among other things, served his distinguished guests like Daniel Webster and Sam Colt the amazing new dessert called "frozen pudding."

The Colonel enjoyed playing the part of a southern planter but underneath he epitomized the sharp Yankee businessman, with a sinister cast. He became a close friend of Henry duPont in Wilmington. Together they rigged production and prices, dominating the powder industry in the United States, and grew rich on sales to the Allies in the Crimean War and to Washington in the Mexican conflict. Like Colt, Hazard was a dyed-in-the-wool Whig who detested abolitionists and other "wicked demagogues" and did a thriving business with the South.

At a meeting in a local tavern his workers, most of whom were English or displaced Irish canal diggers, and undoubtedly duly primed, voted in 1854 to name the village in his honor. In return, their grateful employer erected the Hazard Institute, a combination of library, lecture hall and theatre. With his children Hazard was less fortunate. His favorite, Horace, died in an accidental explosion while testing powder. His other son, a consumptive whom he refused to take into the business, vented his wrath by publicly accusing his father of illegally storing powder and endangering the lives and property of the whole area.

In making black powder, explosions were common occurrences. The materials used were boiled saltpeter from India, sulphur and charcoal, all mixed together with plenty of water in wheel mills. The structures were unheated and unlighted; the powdermen's shoes were tapped with wooden pegs to prevent sparks. Black dust constantly filled the air. Despite all precautions, dozens of employees lost their lives. Their widows received no death benefits, although many were hired to paste labels on powder kegs. Since the railroad refused to transport powder, it had to be carted to the Connecticut River a few miles west, loaded onto flatboats, and at Hartford transferred to coastal schooners.

* * *

In the two decades before 1860 the manufacturing output of the United States more than quadrupled as a truly national market emerged. In 1860, Connecticut ranked fifth. Its total capital investment exceeded $45,000,000, the value of its products $82,000,000. Of course, it had substantially fewer people than the other industrial states, being only one-third the size of Massachusetts, and one-eighth that of New York. But on a per capita basis Connecticut's output of $180 of manufactures placed it second in the country, next to Massachusetts. Despite the trend toward incorporation and the extension of the factory system, partnerships and proprietorships still predominated. Profits in general averaged ten to fifteen percent annually. The population had grown to 460,000—half again the number in 1830 and double that of 1790. New Haven, Hartford, Bridgeport, New London and Waterbury were, in that order, the five largest towns, but none had yet reached a size of 40,000. Of 161,366 persons employed, 42,101 worked on farms and 64,469 in factories. For the first time manufacturing reigned supreme.

Immigration of other than Protestant Anglo-Saxons during the preceding twenty years now showed a definite impact on the makeup of the work force. Over 80,000 foreigners, or one in every six residents, had settled in the state. Nearly three-quarters were Irish Catholics, and many came from Germany. They crowded mostly into the new urban areas like Hartford, New Haven and Norwich, creating the inevitable slum conditions. Even so, only one-third of the population was urbanized. In the opinion of the Congregational clergy, who worried about the decline of Puritan moral and religious values in the villages, the Irish living in the country were considerably more industrious and sober citizens than many reluctant employers (such as Colt) liked to believe. Yet, however rowdy and unskilled the Irish may have been, the steady flow of cheap labor into towns and factories put an end to the shortage of workers which had plagued manufacturers earlier. No longer was it necessary to import skills from abroad. The individual worker's standard of living, however, failed to improve between 1840 and 1860, although the length of his workday gradually shortened from twelve to ten hours.

Neither the influx of foreign-born nor the rival denominations of Baptists, Episcopalians and Methodists shook the supremacy and essential conservatism of Yankee Congregationalism. The Congregational Church, with its spiritual center at Yale, still dominated the religious life of Connecticut. But an interesting development in New Haven signaled the end of the exclusive control of manufacturing by Anglo-Saxons. Three brand new industries were established by recent immigrants, all German Jews. In 1830, Bernard Shoninger started making pianos and organs; Lewis Osterweis, in 1860, pioneered the "Connecticut-type" cigar by blending the Valley broadleaf wrapper and binder with a Cuban filler; and two years later Max Adler and Isaac Strouse introduced sewed corsets.

*   *   *

Historians cite many reasons for the growth of Connecticut manufacturing from the end of the second war with England to the conflict between the states. Commonly mentioned are the Embargo of 1807 and War of 1812; the rising tide of nationalism in New England; the availability of capital and, to some degree, bank credit; the improvements in roads and canals, as well as the coming of the steamboat and railroad; the protective tariff; the accessibility of cheap water-power; and the growing prosperity of the South and West, which created markets for manufactured goods. But often neglected are the boundless energy, amazing skill, outright genius and motivation to succeed of the entrepreneurs themselves, who seized opportunity when it appeared, who never gave in to disappointment or disaster, who always sought new, better, cheaper ways of making more things for more people. For them the race to fortune belonged only to the swift of foot, the stout of heart, the sharp of mind. Lesser mortals did not deserve to survive, at least not as businessmen.

As one awestruck Englishman said after attending the New York Industrial Exposition in 1851: "Certainly the one thing which, more than any other, strikes the visitors to the seats of industrial skill in the United States is the ingenuity, the indomitable energy and perseverance displayed in overcoming the early difficulties which must have stood in the way of anything like successful progress at the outset . . . The early history of nine-tenths of the . . . manufacture now flourishing . . . is that of ruin or of enormous sacrifices on the part of those who had the hardihood to become pioneers . . ."

To bring this industrial saga to a fitting end, let us call the roll of those who in Connecticut left a lasting mark as inventors, manufacturers or salesmen on the development of American technology, business organization and material well-being between the signing of the Constitution and the Civil War:

1740 The Pattisons
1766 Christopher Leffingwell
1775 David Bushnell
1785 John Fitch
1788 Elisha Colt
1788 Jeremiah Wadsworth
1790 Samuel Slater
1793 Eli Terry, Sr.
1795 Simeon North of Berlin
1798 Eli Whitney
1799 The Scholfields
1799 Phineas Pratt
1802 Abel & Levi Porter
1806 The Wilkinsons
1806 David Humphreys
1809 James Brewster
1810 The Hankses
1810 Seth Thomas
1811 Isaac Sanford
1811 The Scovills
1812 Aaron Benedict
1812 The Norths of New Britain
1813 Oliver Wolcott, Jr.
1818 Chauncey Jerome
1819 Edward M. Converse
1823 Ezra Mallory
1826 Samuel Collins
1826 Lambert Hitchcock
1828 Orrin Thompson
1830 Israel Holmes
1831 Dr. John Howe
1832 Elisha K. Root
1833 Robert Wallace
1834 Anson G. Phelps
1835 Frederick T. Stanley
1838 The Cheneys
1839 Joseph Toy

1839 Charles Goodyear
1843 Augustus Hazard
1845 Pliny Jewell
1846 Elias Howe, Jr.
1847 Christopher Spencer
1847 The Rogers Brothers
1848 Almon Farrel
1848 Linus Yale
1848 Samuel Colt
1848 Christian Sharps
1849 William W. Wilcox
1850 Philip Corbin
1852 Horace Wilcox
1856 Charles E. Billings
1857 Wallace Barnes
1858 Eli Whitney Blake
1860 Francis A. Pratt
1860 Amos Whitney
1862 Austin F. Cushman

AND THE YANKEE PEDDLER

Each of these men was responsible for much more than founding a company or inventing a product or a machine. Some, like Whitney, North and Terry, introduced and perfected a complete manufacturing system. Others, like Blake and Goodyear, literally invented industries that never before existed. Still others, such as the Scovill, Cheney and Rogers brothers, by dint of individual effort, a tough moral fibre, the accident of birth and plain luck, laid the foundations for enterprises that developed into the giants of their particular field. The remainder made significant contributions of one kind or another. All were Yankee dreamers and doers.

Thus, it took manufacturing in Connecticut roughly seventy years to come of age, and it remained dominant for a century more. The accomplishments of this economic revolution were many and varied. It led, of course, to the founding of the great machine tool industry in New England, which produced those amazing, self-perpetuating master tools that in turn have made possible all modern industrial production, with its emphasis on uniformity, precision and speed.

For the common man, the miracle of low-cost mass production opened up a Pandora's box of possibilities and possessions. It drastically changed his way of life. It eventually enabled him to buy and enjoy those things that had for centuries been looked upon as luxuries available only to the favored few. It insured the establishment of a solid middle class. In little more than one generation it converted farmer or artisan to wage-earner. It attracted waves of migration from abroad that infused new blood and cultures into a society that had been mostly Anglo-Saxon and Protestant. It created new communities and neighborhoods, as well as turning the larger towns into cities.

Admittedly, the results have not all been good. It can be argued that the machine has dehumanized man because it has become too efficient, complex and all-embracing. Is the role of the urbanite to the conglomerate any better, in human terms, than that of the serf to his feudal lord, the apprentice to his master or the wage-earner to his boss? Have we freed men only to make him a slave to mediocrity, to a rigid code of materialism or to the destruction of his environment? These are questions being asked by the younger generation who seek a simpler, more natural, less fettered life in communes, on farms, anywhere but the cities.

Yet the story of the creation of our industrially-oriented society has relevance for today and tomorrow. This past was nourished on values that will forever endure, if they not be forgotten. The essentials of human motivation and achievement apply to every society that aspires to a better way of life. There can be no real progress without material incentives that fulfill individual desires. Even Marxist dogma in recent years has been compromised to accommodate the capitalist tenets of profit, reward and free choice. Both Marxist and Capitalist might agree, too, that man creates abundance primarily through individual effort and perseverance, in spite of adversity, and frequently it is not so much the personal gain as the striving for it that is ultimately satisfying. This is the real significance of the work ethic, apart from the need to survive. This is what Cotton Mather meant when he enjoined every Christian to have a personal calling "by which his usefulness is distinguished," so that "he may glorify God by doing good for others, and getting of good for himself."

The Twenty-One Leading Manufactures in Connecticut
1860[1]

Industry	Number of Establishments	Number of Hands Employed		Value of Product
		Men	Women	(to nearest 000)
Cotton Goods	129	4,028	4,974	$ 8,911
Woolen Goods	84	2,308	1,459	6,840
Hardware (all types)	118	4,241	505	4,812
Carriages	154	3,313	98	4,172
Hats	53	1,268	519	2,849
Clothing	81	595	4,005	2,758
Paper	55	698	502	2,473
Brass and German Silver	10	897	36	2,334
India Rubber Goods	9	612	197	2,276
Boots and Shoes	212	2,529	777	2,054
Silver-plated & Britannia ware	18	1,013	111	1,959
Flour and meal	132	205	—	1,721
Steam Engines	46	1,189	8	1,711
Hoop skirts	15	426	1,038	1,694
Hosiery	18	481	715	1,384
Sewing silk	19	226	833	1,223
Firearms	15	818	51	1,187
Sewing machines	5	611	—	1,123
Saddlery and Harness	52	743	145	1,121
Clocks	17	896	40	1,085
Gunpowder	4	169	—	1,012
TOTAL OF TWENTY-ONE	1,246	27,266	16,013	$51,850
TOTAL FOR STATE	3,019	44,002	20,467	$81,925

[1] Source: Eighth Census of the United States, 1860.

Connecticut Manufacturing Companies
Founded 1767–1862
Still in Business

Name	Location	Principal Products	Year Founded
Dexter Corporation	Windsor Locks	Paper specialties	1767
Smith-Worthington Saddlery	Hartford	Saddles, bridles and leather goods	1794
Pratt, Read & Co.	Ivoryton	Piano actions, keyboards	1798
Scovill Manufacturing	Waterbury	Primary and fabricated metals	1802
Anaconda-American Brass Co.	Waterbury	Copper and copper alloy mill products	1812
J. M. Ney Company	Bloomfield	Precious metals for dentists and industry	
North & Judd	New Britain	Clothing, industrial and marine hardware	
Waterbury Companies	Waterbury	Plastic parts, buttons and small metal stampings	
Seth Thomas Clocks, Division of General Time	Thomaston	Clocks and metronomes	1813
Manufacturers Association of Connecticut, now Connecticut Business & Industry Association	Hartford	Service to manufacturers	1815
Peter A. Frasse & Co.	Wethersfield	Distributor of steel, aluminum and stainless steel	1816
Gilbert & Bennett Mfg. Co.	Georgetown	Wire fabrics	1818
American Thread Co.	Willimantic	Cotton and worsted yarn and thread	1829
Ingraham Industries, A McGraw-Edison Division	Bristol	Time devices	1831

Name	Location	Principal Products	Year Founded
Belding Heminway Company	Putnam	Silk, cotton, nylon, linen and rayon thread	1832
Bevin Bros. Mfg. Co.	East Hampton	Bells of all types	
Connecticut Printers	Bloomfield	Lithography and letterpress printing	
Neptune Twine & Cord Mills	Moodus	Cotton seine twine	
Rogers Corporation	Rogers	Plastic, rubber and fibrous materials	
American Optical Co.	Putnam	Optical and safety items and scientific instruments	1833
Bigelow Company	New Haven	Steam boilers, heavy fabricated metal parts	
Eagle Lock & Screw Co.	Terryville	Locks, screws, and special hardware	
Russell Manufacturing Div., Fenner-American Ltd.	Middletown	Elastic fabric, friction materials, belts, and belting	1834
Taylor & Fenn Co.	Windsor	Primary and fabricated metals	
Wallace Silversmiths, Div. of Hamilton Watch Co.	Wallingford	Sterling silver, stainless steel flatware and hollowware	1835
E-B Industries	Simsbury	Safety fuses, detonating fuses and chemicals	1836
Farrel Corporation, Div. of U.S.M. Corp.	Ansonia	Heavy machinery	
Bridgeport Fabrics Inc.	Bridgeport	Weatherstripping and auto interior accessories	1837
Cheney Brothers, Inc.	Manchester	Textiles	1838
C. Cowles & Company	New Haven	Automotive hardware	
HCA Industries Inc.	South Norwalk	Men's and women's hats	
Merrow Machine Co.	Hartford	Industrial sewing machines	
Hardware Division of Emhart Corp.	Berlin	Builder's hardware	1839

Name	Location	Principal Products	Year Founded
Sargent & Company	New Haven	Locks, hardware, tools	1840
John McAdams & Sons Inc.	Norwalk	Paper converting machines	1842
American Buckle Co.	West Haven	Wire and sheet metal overall trimmings, and other wire forms	1843
Uni-Royal	Naugatuck	Rubber-soled footwear	
West Haven Buckle Co.	West Haven	Surgical buckles and overhead cable supports	
Miller Company	Meriden	Strip and coil brass and bronze; lighting fixtures	1844
D. & H. Scovill Inc.	Higganum	Eye hoes, harrow discs, abrasive adhesives	
S. Curtis & Sons Inc.	Sandy Hook	Folding paper boxes	1845
International Silver Co.	Meriden	Sterling and plate silverware	1847
New Haven Gas Company	New Haven	Manufactured gas	
M. S. Brooks & Sons, Inc.	Chester	Wire goods	1848
Colt's Patent Fire Arms, Div. of Colt Industries	Hartford	Small arms	
Connecticut Natural Gas	Hartford	Gas distribution	
McLagon Foundry Co.	New Haven	Gray iron and semi-steel castings	
Davis-Standard Division, Crompton-Knowles	Pawcatuck	Extruding machinery for plastic and rubber	1848
Turner & Seymour Mfg. Co.	Torrington	Gray iron castings and metal fabrications	
Blake & Johnson Company	Waterville	Screw machine products	1849
Southern Conn. Gas Co.	Bridgeport	Manufactured gas	

Name	Location	Principal Products	Year Founded
Hoggson & Pettis Mfg. Co.	Wallingford	Small tools, steel molds and dies, marking devices	
Ball & Socket Mfg. Co.	Cheshire	Buttons, bells, metal stampings	1850
Bristol Brass Corp.	Bristol	Sheet, rod and wire brass	
Robertson Paper Box Co.	Montville	Paperboard and folding paper boxes	
Sheffield Tube Corp.	New London	Toothpaste and cosmetic tubes	
Singer Company	Bridgeport	Sewing machines and military products	
Valve & Instrument Div., Dresser Industries (formerly Manning, Maxwell & Moore Inc.)	Stratford	Gages, valves and instruments	1851
Merriam Mfg. Co.	Durham	Metal boxes and displays	
Waterbury Farrel Foundry, Div. of Textron Inc.	Cheshire	Metalworking machinery	
Stanley Works	New Britain	Builders' hardware, tools, wire goods, metal products	1852
Waterbury Buckle Co.	Waterbury	Sheet metal and wire specialties	1853
Wauregan Mills Inc.	Wauregan	Cotton and synthetic fabrics	
Brunswick Corp.	Torrington	Skates, golf shafts, fishing rods, athletic shoes	1854
Clark Bros. Bolt Co.	Milldale	Bolts, nuts, screws, rivets	
Kerite Company	Seymour	Insulated wire and cable	
Herman Roser & Sons Inc.	Glastonbury	Pigskin leather	
J. H. Sessions & Son	Bristol	Metal stampings	

Name	Location	Principal Products	Year Founded
I. L. Stiles & Son Brick Company	North Haven	Brick and cinder blocks	
Cottrell Company, Div. of Harris-Intertype Corp.	Pawcatuck	Printing presses	1855
Alloy Foundries Division, Eastern Company	Naugatuck	Castings and nonferrous hardware	1856
Glenbrook Laboratories, Div. of Sterling Drug Inc.	Glenbrook	Pharmaceuticals, dentrifices and cosmetics	
Turner & Stanton Co.	Norwich	Braided and twisted cotton cord	
Whiton Machine Company	New London	Lathe chucks, steam turbines	
Wallace Barnes Division, Associated Spring Corp.	Bristol	Precision mechanical springs	1857
Bridgeport Hydraulic Co.	Bridgeport	Public water supply	
I. S. Spencers Sons Inc.	Guilford	Jobbing foundry and machine shop	
Timex	Waterbury	Watches and clocks	
J. T. Henry Company	Hamden	Shears and pneumatic pruners	1859
Pratt & Whitney Machine Tool Div., Colt Industries	West Hartford	Machine tools, cutting tools and gages	1860
Case Brothers Inc.	Manchester	Specialty paper board	1861
Strouse, Adler Company	New Haven	Foundation garments and paper boxes	
Cushman Industries	Hartford	Chucks, air cylinders, power wrenches	1862

Acknowledgment

THE major justification for this book is that no primarily social history of early manufacturing in Connecticut has been published that attempts to integrate the entire subject for the layman. There are a few excellent accounts of individual companies, of inventors like Eli Whitney, of personalities like P. T. Barnum, and even of entire industries like tinsmithing, silverplating and carpetmaking. These are all listed in the References.

My approach has, purposefully, been selective. I did not wish to bore the reader with a chronological or statistical record. I have tried to include only material that was interesting or significant. I apologize to those companies which may have received short shrift or no mention at all, simply because of lack of space.

Those wishing to probe more deeply into a particular phase of this period should find one or more leads in the References, which is as complete as I could make it.

Finally, let me express my gratitude to Dr. Glenn Weaver, Professor of History at Trinity College, and to Dr. D. G. Brinton Thompson, Professor Emeritus of History at Trinity, for reading the manuscript and offering valuable suggestions; to Thompson R. Harlow, Melancthon W. Jacobus and Doris E. Cook of the Connecticut Historical Society for their unstinted assistance; to Elizabeth Oliver for her keen editorial eye; and to my secretary Patricia Pratt, whose patience in typing and retyping the manuscript never gave out.

ELLSWORTH S. GRANT

West Hartford, Connecticut
June 1, 1974

References

Abbe, Nellie G. "Traffic on the Connecticut River", in *Connecticut Quarterly*, Vol. III, Jul.-Sept. 1897

Alden, J. Deane *Proceedings at the Dedication of Charter Oak Hall*, Hartford: Case, Tiffany and Co., 1856

Allen, Edward "Powder Hollow", in *Connecticut Circle*, Sept. and Oct. 1947

Allis, Marguerite *Connecticut River*, New York: Putnam, 1939

Andrews, Charles M. "On Some Early Aspects of Connecticut History", in *New England Quarterly*, Vol. 17, No. 1, Mar. 1944

The River Towns of Connecticut, Johns Hopkins University, *Studies in History and Political Science*, 7th Series, Baltimore, 1889

Bacon, Edwin M. *The Connecticut River*, New York and Lond.: G. P. Putnam's Sons, 1911

Barber, John W. *Connecticut Historical Collections*, New Haven: 1836

Barnard, Henry, ed. *Armsmear . . . a Memorial*, New York: Alvord, 1864

Barnes, Carlyle F. *Associated Spring Corporation*, New York: Newcomen Society in North America, 1963

Barnum, Phineas T. *Struggles and Triumphs; or, Forty Years' Recollections*, New York: American News Co., 1871

Barr, Lockwood "Joseph Ives' Wagon Spring Clocks", in *The Chronicle of the Early American Industries Association*, Vol. II, No. 24, Sept. 1943

Beals, Carleton *Our Yankee Heritage*, New York: D. McKay Co., 1955

Bidwell, Percy W. "Rural Economy in New England at the Beginning of the Nineteenth Century", *Transactions* of the Conn. Academy of Arts and Sciences, Vol. XX, Apr. 1916, New Haven

Bingham, Leslie M. "Builders' Hardware", in *Connecticut Industry*, Vol. 12, No. 10, Oct. 1934

"Machine Tools", in *Connecticut Industry*, Vol. 13, No. 5, May 1935

Bishop, James L. *A History of American Manufactures from 1608–1860*, Philadelphia: Edward Young & Co., 1864

Blackall, Frederick S. *Invention and Industry—Cradled in New England!* New York: Newcomen Society of England, American Branch, 1946

Blake, William P. *History of the Town of Hamden*, New Haven: Price & Lee, 1888
    "Sketch of the Life of Eli Whitney", *Papers of the New Haven Colony Historical Society*, Vol. V, No. 4, New Haven, 1894

Bolles, Albert S. *Industrial History of the United States*, Norwich, Conn.: Bill Pub. Co., 1879

Bowen, Catherine Drinker *Miracle at Philadelphia*, Boston: Little, Brown, 1966

Brett, John A. *Connecticut Yesterday and Today*, Hartford: The John Brett Co., 1936

Brewer, Thomas B. *The Formative Period of 140 American Manufacturing Companies, 1789–1929*, unpublished Ph. D. dissertation, University of Pennsylvania, 1962

Brookes, George S. *Cascades and Courage*, Rockville, Conn.: T. F. Rady & Co., 1955

Bruchey, Stuart W. *The Roots of American Economic Growth, 1607–1861*, Harper & Row, 1968

Buell, Carleton W. & Lockwood Barr *Clockmakers of Bristol, Connecticut*, unpublished manuscript in 2 vols., June, 1937, Yale University Library, New Haven

Case, Anne B. "The Collins Company of Collinsville", *The Lure of the Litchfield Hills*. Vol. 8, Jun. 1944

Chamberlain, John *The Enterprising Americans*, New York: Harper, 1963

Chandler, John B. "Industrial History", in *History of Connecticut*, Vol. IV, Norris G. Osborn, ed., New York: The States History Co., 1925

Clark, George L. *A History of Connecticut*. New York and Lond.: Putnam's, 1914

Clark, Victor S. *History of Manufactures in the United States, 1607–1914*, New York: McGraw-Hill, 1929

Coffin, David L. *The History of the Dexter Corporation*, New York: The Newcomen Society in North America, 1967

Cole, Arthur H. *The American Carpet Manufacture*, Cambridge: Harvard University Press, 1941
    The American Wool Manufacture, Cambridge: Harvard University Press, 1926

Collins, Samuel W. *The Collins Company, 1826–1867: Reminiscences of Samuel Watkinson Collins*. Typescript at The Connecticut Historical Society.

The Collins Company *One Hundred Years*, Springfield: Bible-Plimpton Co., 1926

Colt, Samuel Letters and Papers in The Connecticut Historical Society

Colt, Samuel *Saml. Colt's Own Record . . . 1847*, Hartford: The Connecticut Historical Society, 1949

Connecticut Society for the Encouragement of American Manufactures *Address*, Middletown: T. Dunning, 1817

Crocker, Antoinette Cheney, *Letters of Frank Woodbridge Cheney*

Cross, Wilbur L. *Connecticut Yankee*, New Haven: Yale University Press, 1943

Davis, Joseph S. *Eighteenth Century Business Corporations in the United States*, Vol. II of *Essays in the Earlier History of American Corporations*, Cambridge: Harvard University Press, 1917

Davis, William T., ed. *The New England States*, Boston: D. H. Hurd & Co., 1897

Day, Clive *The Rise of Manufacturing in Connecticut, 1820–1850*. New Haven, Pub. for the Tercentenary Commission by the Yale University Press, 1935

DeVoe, Shirley S. *The Tinsmiths of Connecticut*, Middletown: Wesleyan University Press, 1968

Deyrup, Felicia J. *Arms Makers of the Connecticut Valley* . . . *1798–1870*, Smith College Studies in History, Northampton, Mass., 1948

Dolan, J. R. *The Yankee Peddlers of Early America*. New York: Bramhall House, 1964

Drepperd, Carl W. *American Clocks and Clockmakers*, Boston: C. T. Branford Co., 1958

Durant, Will and Ariel *The Lessons of History*, New York: Simon and Schuster, 1968

Dwight, Timothy *Travels in New-England and New York*, New Haven: T. Dwight, 1821–22

Earle, Alice M. "Gallant Silken Trade", in *New England Magazine*. Jul. 1900

Edwards, William B. *Story of Colt's Revolver*, Harrisburg, Pa.: Stackpole Co., 1953

Ellsworth, John E. *Ensign-Bickford Company and the Safety Fuse in America*, Chicago: R. R. Donnelley Sons & Co., 1936

Simsbury . . . *1642–1935*, Simsbury: Simsbury Committee for the Tercentenary, 1935

Erving, Henry W. *The Connecticut River Banking Company, 1825–1925*, Hartford, 1925

Evans, C. H., Jr. *Business Incorporations in the United States, 1800–1943*, New York: National Bureau of Economic Research *Publications*, No. 49, 1948

Ewing, John S. and Norton, Nancy F. *Broadlooms and Businessmen*, Cambridge: Harvard University Press, 1955

Fuller, Grace P. *An Introduction to the History of Connecticut as a Manufacturing State*, Smith College Studies in History, Northampton, Mass., 1915

Glover, John G. and Cornell, William B. *The Development of American Industries*, New York: Prentice-Hall, Inc., Rev. ed., 1941

Goodrich, Samuel G. *Down By the old Mill Stream*

Goodwin, Francis II "Some Highlights on the Maritime History of the Connecticut River". Unpublished manuscript

Greeley, Horace *et al. The Great Industries of the United States*, Hartford: Burr & Hyde, 1872

Green, Constance M. *Eli Whitney and the Birth of American Technology*, Boston: Little, Brown, 1956

Hamilton, Alexander *Industrial and Commercial Correspondence*, ed. by Arthur Cole, Chicago: A. W. Shaw Co., 1928

Hard, Walter *The Connecticut*, New York, Toronto: Rinehart & Co., 1947

Hart, Samuel *In Memoriam Samuel Colt and Caldwell Hart Colt*, Springfield: Clifton Johnson, 1898

Harte, Charles R. *Connecticut's Canals*. Reprinted from the 54th Annual Report of the Connecticut Society of Civil Engineers, 1938

Haven, Charles T. and Belden, Frank A. *A History of the Colt Revolver*, New York: W. Morrow & Co., 1940

Hoopes, Penrose R. *Early Clockmaking in Connecticut*, New Haven, Pub. for the Tercentenary Commission by the Yale University Press, 1934

Hubbard, Guy "Development of Machine Tools in New England", in *American Machinist*, Vols. 59–61, 1923–24

Humphreys, David *The Miscellaneous Works of David Humphreys*, New York: T. and J. Swords, 1804

Humphreys, Frank L. *Life and Times of David Humphreys*, New York and Lond.: Putnam, 1917

Jacobus, Melancthon W. *The Connecticut River Steamboat Story*, Hartford: The Connecticut Historical Society, 1956

Jefferson, Thomas *Writings*, Washington, D.C.: Thomas Jefferson Memorial Association of the United States, 1905

Jerome, Chauncey *History of the American Clock Business for the Past Sixty Years*, New Haven: F. C. Dayton, Jr., 1860

Johnston, Alexander *Connecticut*, American Commonwealths Series, Boston & New York: Houghton, Mifflin & Co., 1900

Keir, Robert Malcolm *The Epic of Industry*, New Haven: Yale University Press, 1926
Manufacturing Industries in America, New York: Ronald Press, 1928
"The Unappreciated Tin Peddler" *Annals* of the American Academy of Political and Social Science, Vol. 46, Mar. 1913

Kenney, John T. "Lambert Hitchcock of Hitchcocksville, Connecticut", in *The Connecticut Antiquarian*, Vol. 18, Jul. 1966

Kihn, Phyllis "Colt in Hartford", in *The Connecticut Historical Society Bulletin*, Jul. 1959

Kimball, Milo *Principles of Corporate Finance*, New York, Lond.: Longmans, Green and Co., 1957

Kirkland, Edward C. *Men, Cities, and Transportation*, Cambridge: Harvard University Press, 1948

Larned, Ellen D. *History of Windham County, Connecticut*, Vol. II, Worcester, Mass.: C. Hamilton, 1880

Lathrop, William G. *The Development of the Brass Industry in Connecticut*, New Haven, Pub. for the Tercentenary Commission by the Yale University Press, 1936

Lee, William S. *The Yankees of Connecticut*, New York: Henry Holt & Co., 1957

Marburg, Theodore F. *Management Problems and Procedures of a Manufacturing Enterprise 1802–52, A Case Study of the Origin of the Scovill Manufacturing Company*, Ph. D. dissertation, Clark University, Worcester, Mass. 1942

March, Daniel *Yankee Land and the Yankee*, Hartford: Case, Tiffany and Burnham, 1840

Martin, Margaret E. *Merchants and Trade of the Connecticut River Valley, 1750–1820*, Department of History of Smith College, Northampton, Mass., 1939

Marx, Leo *The Machine in the Garden*, New York: Oxford University Press, 1964

Mather, Cotton *Two Brief Discourses . . . A Christian in . . . his Personal Calling . . .* Boston: B. Green & J. Allen, 1701. In *Early American Imprints 1639–1800*, No. 990. Worcester: American Antiquarian Society (from the copy at Yale University)

May, Earl C. *Century of Silver, 1847–1947*, New York: R. M. McBride & Co., 1947

Meriden, Conn. Sesquicentennial Committee *100 Years of Meriden*. Meriden, 1956

Middlebrook, Louis F. *History of Maritime Connecticut during the American Revolution, 1775–1783*, Salem: The Essex Institute, 1925

Mirsky, Jeannette and Nevins, Allan *The World of Eli Whitney*, New York: Macmillan, 1952

Mitchell, James L. *Colt, A Collection of Letters and Photographs,* Harrisburg, Pa.: Stackpole Co., 1959

Moore, Mabel R. *Hitchcock Chairs*, New Haven, Pub. for the Tercentenary Commission by the Yale University Press, 1933

Morgan, Forrest, ed. *Connecticut as a Colony and as a State*, Hartford: Publishing Society of Connecticut, 1904

Morison, Samuel Eliot *The Oxford History of the American People*, Rev. ed., New York: Oxford University Press, 1965

Morse, Jarvis Means *A Neglected Period of Connecticut's History, 1818–1850*, New Haven: Yale University Press, 1933

Mussey, June Barrows *Young Father Time, a Yankee Portrait*, New York: Newcomen Society in North America, 1950

Nelson, Henry L. "The Cheneys' Village at South Manchester, Conn.", in *Harper's Weekly*, Feb. 1, 1890

Niven, John *Connecticut for the Union*, New Haven: Yale University Press, 1965

North, Douglass C. *The Economic Growth of the United States, 1790–1860* New York: W. W. Norton & Co., 1966

North, S. N. D. and Ralph H. *Simeon North*, Concord, N.H.: Rumford Press, 1913

Nye, Russell B. *This Almost Chosen People*, East Lansing, Mich.: Michigan State University Press, 1966

Olmstead, D. *Memoir of Eli Whitney*, New Haven: Durrie & Peck, 1846

Parrington, Vernon L. *Main Currents in American Thought*, New York: Harcourt, Brace, 1927

Pease, John C. and Niles, John M. *A Gazetteer of the States of Connecticut & Rhode Island*, Hartford: William S. Marsh, 1819

Peirce, Bradford K. *Trials of an Inventor: Life & Discoveries of Charles Goodyear*, New York: Carleton & Porter, 1866

Pierson, George W. *Tocqueville and Beaumont in America*, New York: Oxford University Press, 1938

Prentice, Archibald *Tour of the United States*, London: John Johnson, 1849

Purcell, Richard J. *Connecticut in Transition, 1775–1818*, Washington: American Historical Association, 1918

Rand, Christopher *The Changing Landscape: Salisbury, Connecticut*, New York: Oxford University Press, 1968

Roberts, George S. *Historic Towns of the Connecticut River Valley*, Schenectady: Robson & Adee, 1906

Roe, Joseph W. *Connecticut Inventors*, New Haven, Pub. for the Tercentenary Commission by the Yale University Press, 1934
English and American Tool Builders, New York: McGraw-Hill, 1926

"Inventors and Engineers of Old New Haven", New Haven: New Haven Colony Historical Society, 1939

Rohan, Jack *Yankee Arms Maker*, New York and London: Harper & Bros., 1935

Schlesinger, Arthur M., Jr. *The Age of Jackson*, Boston: Little, Brown, 1945

Shepard, Odell *Pedlar's Progress; the Life of Bronson Alcott*, Boston: Little, Brown, 1937

Sterry, Iveagh H. and Garrigus, William H. *They Found a Way*, Brattleboro, Vt.: Stephen Daye Press, 1938

Stewart, George R. *Names on the Land*, New York: Random House, 1945

Taylor, George R. "The Transportation Revolution, 1815–60", *The Economic History of the United States*, Vol. IV, New York: Rinehart, 1951

Terry, Henry *American Clock Making*, Waterbury: J. Giles & Son, 1870

Tocqueville, Alexis de *Democracy in America*, New York, Alfred A. Knopf, 1945

Trumbull, Benjamin *A Complete History of Connecticut*. New Haven: Maltby, Goldsmith & Co., 1818

Trumbull, James H. *The Public Records of the Colony of Connecticut . . . 1636–1665*, Hartford: Brown & Parsons, 1850

U.S. Treasury Dept. *Report of the Secretary . . . on the Subject of Manufactures*, Philadelphia: Childs & Swaine, 1791
A Statement of the Arts & Manufactures of the United States, 1810 . . . prepared by Tench Coxe, Philadelphia: A. Cornman, Jr., 1814

Van Dusen, Albert E. *Connecticut*, New York: Random House, 1961

Verrill, A. Hyatt *The Heart of Old New England*, New York: Dodd, Mead & Co., 1936

Wallace, Irving *The Fabulous Showman*, New York: Alfred A. Knopf, 1959

Ware, Caroline F. *The Early New England Cotton Manufacture*, Boston and New York: Houghton Mifflin, 1931

Ware, Norman *The Industrial Worker, 1840–1860*, Boston and New York: Houghton Mifflin, 1924

Washington, George *Diaries . . .* Ed. John C. Fitzpatrick, Boston and New York: Houghton Mifflin, 1925

Welles, Arnold "Father of our Factory System", in *American Heritage*, Vol. IX, No. 3, Apr. 1958

White, George S. *Memoir of Samuel Slater*, Phila., 1836

Wilcox, Crittenden & Co. *Century of Dependability 1847*, Middletown: Wilcox, Crittenden & Co., 1947

Wright, Carroll D. "Report on the Factory System of the United States", in *Report on the Manufactures of the United States at the Tenth Census*, Tenth Census Vol. X, Washington: Govt. Printing Office, 1883

Wright, Richardson *Hawkers and Walkers in Early America*, Phila.: J. B. Lippincott Co., 1927

MISCELLANEOUS SOURCES

County Histories
Town Histories
Connecticut Courant
Hartford Courant
Hartford Daily Post
Hartford Evening Press
U.S. Census Reports 1850, 1860, 1880
Diaries, Letters, Memoirs

Illustration Credits

* Connecticut Historical Society

Index

Mt. Riga ● State Park

● Robertsville
● Riverton
Lakeville ● ● Amesville
● Grantville Tar
● Huntsville ● Mooreville
● Burrville
Wrightville ● ● Bakersville
Torrington ■ ● Collins
Unionville ●
Whigville ●
● Terryville ● F
Thomaston ■ ■ Plymouth ■ Bristol
● Merwinville
● Gaylordsville
● Northville Waterville ● ● Plant
Hotchkissville ● ■ M
● Wellsville Bradleyville ● ■ Waterbury
● Mixville
Cheshire ● ■
Lanesville ● ■ Naugatuck

Hamden ■
Augerville ● ■ No
■ Danbury ■ Seymour
Whitneyville ●
■ Ansonia
Westville ● ■ New H
West ■
Haven

Stratford ■
Bridgeport ■

MAJOR MANUFACTURING T

Norwalk ■
South Norwalk ■

● Glenville ■ TOWNS
● VILLAGES